Getting Ready for the PSSA

Grade 2

Houghton Mifflin Harcourt

INCLUDES:
- Beginning-of-Year, Middle-of-Year, and End-of-Year Benchmark Tests
- Chapter Tests
- All Assessments in the Format of the PSSA
- Individual Record Forms

Copyright © by Houghton Mifflin Harcourt Publishing Company

All rights reserved. No part of this work may be reproduced or transmitted in any form or by any means, electronic or mechanical, including photocopying or recording, or by any information storage and retrieval system, without the prior written permission of the copyright owner unless such copying is expressly permitted by federal copyright law. Requests for permission to make copies of any part of the work should be submitted through our Permissions website at https://customercare.hmhco.com/permission/Permissions.html or mailed to Houghton Mifflin Harcourt Publishing Company, Attn: Intellectual Property Licensing, 9400 Southpark Center Loop, Orlando, Florida 32819-8647.

Printed in the U.S.A.

ISBN 978-0-544-90006-6

2 3 4 5 6 7 8 9 10   2266   25 24 23 22 21 20 19 18 17 16

4500614887   A B C D E F G

If you have received these materials as examination copies free of charge, Houghton Mifflin Harcourt Publishing Company retains title to the materials and they may not be resold. Resale of examination copies is strictly prohibited.

Possession of this publication in print format does not entitle users to convert this publication, or any portion of it, into electronic format.

Contents

Overview of Go Math! PSSA Assessment Guide .... **v**

PSSA Assessment Formats .... **vii**

Management Forms

Individual Record Forms **AG125**

Tests and Management Forms

Beginning-of-Year Test AG1
Middle-of-Year Test AG13
End-of-Year Test AG25
Individual Record Form AG125

Chapter 1
Chapter 1 Test AG37
Individual Record Form AG128

Chapter 2
Chapter 2 Test AG45
Individual Record Form AG129

Chapter 3
Chapter 3 Test AG53
Individual Record Form AG130

Chapter 4
Chapter 4 Test AG61
Individual Record Form AG131

Chapter 5
Chapter 5 Test AG69
Individual Record Form AG132

Chapter 6
Chapter 6 Test AG77
Individual Record Form AG133

Chapter 7
Chapter 7 Test AG85
Individual Record Form AG134

Chapter 8
Chapter 8 Test AG93
Individual Record Form AG135

Chapter 9
Chapter 9 Test AG101
Individual Record Form AG136

Chapter 10
Chapter 10 Test AG109
Individual Record Form AG137

Chapter 11
Chapter 11 Test AG117
Individual Record Form AG138

Overview of *Go Math!* PSSA Assessment

How Assessment Can Help Individualize Instruction

The PSSA *Assessment Guide* contains several types of assessment for use throughout the school year. The following pages will explain how these assessments can be utilized to diagnose children's understanding of the Pennsylvania Common Core Standards and to guide instructional choices, improve children's performance, and help facilitate children's mastery of this year's objectives.

Diagnostic Assessment

Beginning-of-Year Test in the PSSA *Assessment Guide* is a mixed-response-format test comprised of multiple-choice and open-ended items. This test should be utilized early in the year to establish on-grade-level skills that children may already understand. This benchmark test will allow customization of instructional content to optimize the time spent teaching specific objectives. Suggestions for intervention are provided for this test.

Formative Assessment

Middle-of-Year Test in the PSSA *Assessment Guide* assesses the same skills as the Beginning-of-Year Test, allowing monitoring of children's progress to permit instructional adjustments when required.

Summative Assessment

Chapter Test in the PSSA *Assessment Guide* measures children's mastery of concepts and skills taught in the chapter. The test assesses the mastery of the Pennsylvania Common Core Standards taught in the chapter. It is a mixed-response-format test comprised of multiple-choice and open-ended items.

End-of-Year Test in the PSSA *Assessment Guide* documents each child's level of mastery of the concepts and skills for the current grade level and mirrors the Beginning- and Middle-of-Year Tests. Used together, these mixed-response-format tests allow for monitoring of growth throughout the year.

PSSA Assessment Formats

The state of Pennsylvania has developed assessments that contain mixed responses comprised of multiple-choice and open-ended items. This allows for a more robust assessment of students' understanding of concepts. PSSA assessments will be administered via computers and *Go Math!* presents items in formats similar to what children will see on the tests. The following information is provided to help teachers familiarize children with these different types of items. An example of each item type appears on the following pages. You may want to use the examples to introduce the item types to children. The following explanations are provided to guide children in answering the questions.

Example 1 Divide a two-digit number by a one-digit number.

Multiple Choice

For this type of item, children respond to a single question with several choices. There will be a question and children will choose the correct choice. It is important for children to know they must fill in the bubble to make their choice. There will only be one choice that is correct.

Example 2 Solve and explain a problem with multiple parts.

Open-ended

This type of item will ask children to solve a problem that has multiple parts. The children will be expected to show their work. This can be shown with an explanation or with a drawing. Tell them to think about their answers carefully. There are some questions that will have more than one correct answer.

1. Which of these is a way to show 42?

   ○ 4 + 2   ○ [base-ten blocks]   ○ 40 + 20   ○ ||||

2. Jill and Ed collect postcards. Jill has 124 postcards. Ed has 131 postcards.

   A. Who has more postcards?

   PUT your answer in the BLANK BELOW.

   Answer: _____

   B. Jill gets 10 more postcards. Ed gets 5 more postcards. Who has more postcards now?

   SHOW all your WORK.

   Answer: _____

   C. The next week, Jill gets 18 more postcards. Ed gets 15 more postcards. Who has more postcards now?

   SHOW all your WORK.

   Answer: _____

Name _____

Beginning-of-Year Test
Page 1

Choose the correct answer.

1. Which shows a related addition fact?

    $15 - 8 = 7$

    ○ $15 + 7 = 22$
    ○ $8 - 7 = 1$
    ○ $7 + 8 = 15$
    ○ $23 - 8 = 15$

2. There are 9 bugs on the grass and 5 bugs on a leaf. Which number sentence shows how many bugs there are in all?

    ○ $10 + 5 = 15$
    ○ $9 + 5 = 14$
    ○ $9 - 5 = 4$
    ○ $5 + 4 = 9$

3. Gina has 4 green trains, 2 red trains, and 6 yellow trains. How many trains does Gina have in all?

    ○ 6
    ○ 8
    ○ 10
    ○ 12

4. There are 725 students in the school. There are 343 boys. How many girls are there?

Hundreds	Tens	Ones
☐	☐	☐
7	2	5
− 3	4	3

    ○ 382   ○ 422
    ○ 428   ○ 482

GO ON

Assessment Guide
© Houghton Mifflin Harcourt Publishing Company

AG1

Beginning-of-Year Test

Name _____

Beginning-of-Year Test
Page 2

5. Chelsea's soccer team collected 427 cans for donation. Jordan's basketball team collected 378 cans. How many cans did the two teams collect?

  ○ 685   ○ 713
  ○ 805   ○ 937

6. What is the difference?

$$\begin{array}{r} 402 \\ -\ 173 \\ \hline \end{array}$$

  ○ 229   ○ 329
  ○ 331   ○ 339

7. Use an inch ruler. What is the length of the ribbon to the nearest inch?

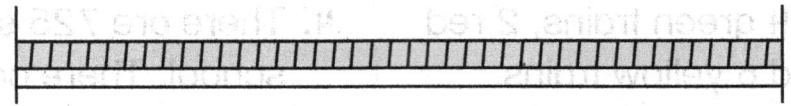

  ○ 2 inches   ○ 4 inches
  ○ 6 inches   ○ 8 inches

8. Glen made a line plot to show the lengths of his toy airplanes. How many planes are shown in the line plot?

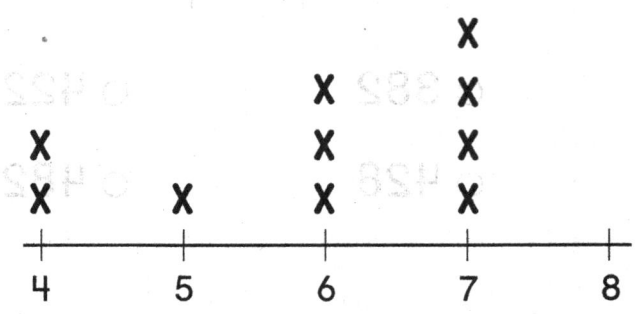

Lengths of Toy Airplanes in Inches

  ○ 7
  ○ 8
  ○ 9
  ○ 10

GO ON

Assessment Guide
© Houghton Mifflin Harcourt Publishing Company

AG2

Beginning-of-Year Test

Name _____

Beginning-of-Year Test
Page 3

9. Which is the **best** estimate of the length of a baseball bat?

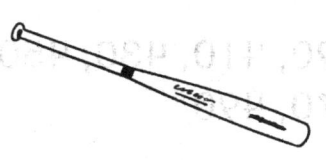

- ○ 2 feet
- ○ 6 feet
- ○ 8 feet
- ○ 10 feet

10. Fred wants to measure the distance around a ball. Which is the **best** tool for Fred to use?

- ○ counters
- ○ cup
- ○ measuring tape
- ○ pencil

11. Ms. Angeles writes an odd number on the board. Which could be the number that Ms. Angeles writes?

- ○ 3
- ○ 4
- ○ 6
- ○ 8

12. What is the value of the underlined digit?

<u>3</u>8

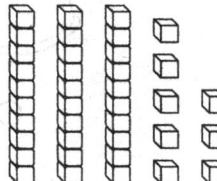

- ○ 3
- ○ 8
- ○ 30
- ○ 80

Assessment Guide
© Houghton Mifflin Harcourt Publishing Company

AG3

Beginning-of-Year Test

Name _____

Beginning-of-Year Test
Page 4

13. Which shows another way to write the number?

57

○ 7 tens 5 ones

○ fifty-seven

○ 5 + 70

○ 5 + 7

14. Kayleigh starts at 370 and counts by tens. What are the next 6 numbers Kayleigh will say?

○ 390, 410, 430, 450, 470, 490

○ 380, 390, 400, 410, 420, 430

○ 375, 385, 395, 405, 415, 425

○ 371, 372, 373, 374, 375, 376

15. Which object is shaped like a cylinder?

 ○

 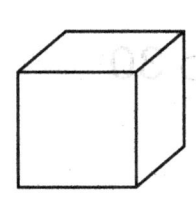
 ○

16. Danny makes a pentagon with straws. He uses one straw for each side of the shape. How many straws does Danny need?

○ 1

○ 3

○ 4

○ 5

GO ON

Assessment Guide
© Houghton Mifflin Harcourt Publishing Company

AG4

Beginning-of-Year Test

Name _____

Beginning-of-Year Test
Page 5

17. Which of these shapes has **fewer** than 4 angles?

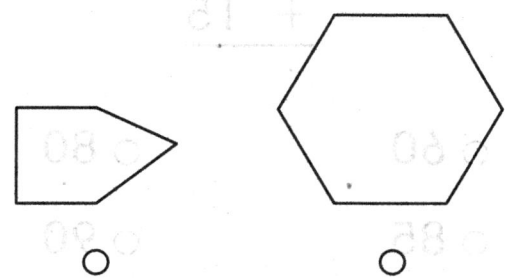
○   ○

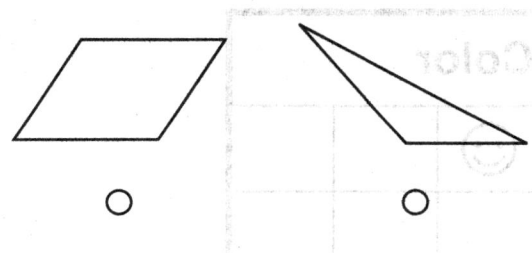
○   ○

18. Which shows a half of the shape shaded?

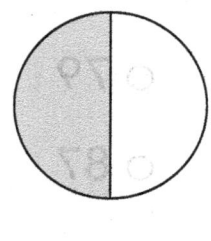 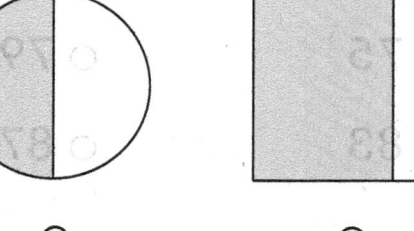
○   ○

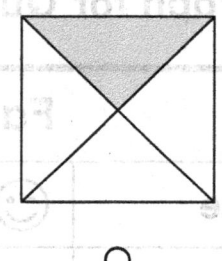 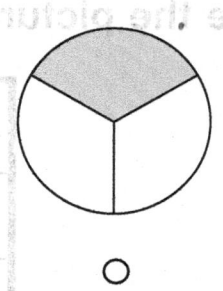
○   ○

19. A store sold 21 green rings and 38 red rings. Which number sentence tells how many rings the store sold?

○ 38 − 21 = 17

○ 38 + 0 = 38

○ 21 + 38 = 59

○ 12 + 38 = 50

20. What is the sum?

$$\begin{array}{r} 24 \\ 15 \\ +\ 36 \\ \hline \end{array}$$

○ 39

○ 51

○ 65

○ 75

Assessment Guide
© Houghton Mifflin Harcourt Publishing Company

AG5

Beginning-of-Year Test

Name _____

Beginning-of-Year Test
Page 6

21. Jared has 29 blocks. Amanda has 46 blocks. How many blocks do they have?

○ 75   ○ 79
○ 83   ○ 87

22. What is the sum?

$$\begin{array}{r} 75 \\ +\ 15 \\ \hline \end{array}$$

○ 60   ○ 80
○ 85   ○ 90

Use the picture graph for Questions 23–24.

Favorite Color					
blue	☺	☺	☺		
green	☺	☺			
red	☺	☺	☺	☺	☺

Key: Each ☺ stands for 1 child.

23. How many **more** children chose red than green?

○ 2   ○ 3
○ 5   ○ 7

24. **2 more** children choose green. How many ☺ should be in the green row now?

○ 3   ○ 4
○ 5   ○ 7

GO ON

Name _____

Beginning-of-Year Test
Page 7

25. Ian made a tally chart of the flowers he planted.

Flowers Planted	
Flower	Tally
roses	||||
daisies	⋕|||
tulips	⋕

How many tulips did Ian plant?

○ 4 ○ 5
○ 6 ○ 8

26. Use the bar graph.

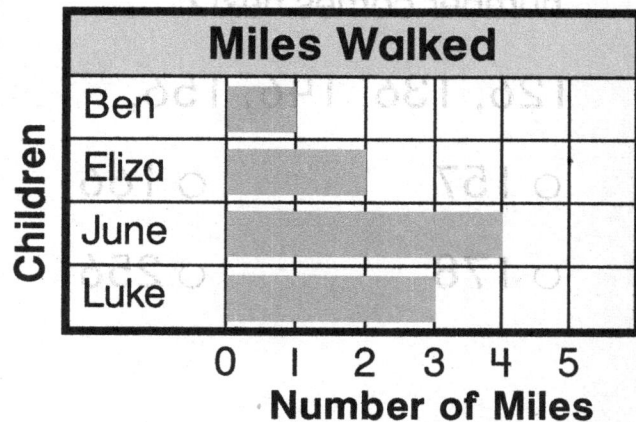

Who walked the **fewest** number of miles?

○ Ben ○ June
○ Eliza ○ Luke

27. Which number has the digit 4 in the hundreds place?

○ 42
○ 140
○ 453
○ 964

28. Which shows another way to write the number?

five hundred thirty-seven

○ 5 + 3 + 7
○ 50 + 37
○ 50 + 30 + 7
○ 500 + 30 + 7

GO ON

Assessment Guide
© Houghton Mifflin Harcourt Publishing Company

AG7

Beginning-of-Year Test

Name _____

Beginning-of-Year Test
Page 8

29. Look at the pattern. What number comes next?

126, 136, 146, 156, ▨

○ 157   ○ 166
○ 178   ○ 256

30. Which comparison is **true**?

○ 542 > 621
○ 382 > 405
○ 261 < 243
○ 215 < 225

31. Melissa gave her brother these coins. What is the value of the coins?

30 cents   65 cents   80 cents   95 cents
   ○   ○   ○   ○

32. Lara wants to buy a marker that costs one dollar. Which coins have a total value of one dollar?

○ 100 dimes

○ 100 pennies

○ 10 pennies

○ 10 nickels

GO ON ▶

Assessment Guide
© Houghton Mifflin Harcourt Publishing Company

AG8

Beginning-of-Year Test

Name _____

Beginning-of-Year Test
Page 9

33. Jason went on a morning run at the time shown on the clock. At what time did Jason go for his run?

9:55 A.M.   10:45 A.M.   9:55 P.M.   10:45 P.M.
  ○   ○   ○   ○

---

34. Break apart the ones to subtract. What is the difference?

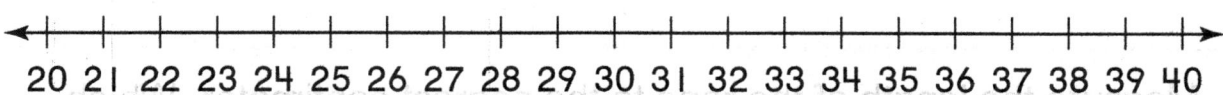

33 − 6 = _____

27   28   29   30
○   ○   ○   ○

---

35. Rita had 37 pencils. She gave away 14 pencils. Which number sentence can be used to find how many pencils Rita has now?

○ 23 − ☐ = 14

○ 30 + ☐ = 37

○ 37 − 14 = ☐

○ 37 + 14 = ☐

36. Steven picks 22 berries. He picks 18 more berries. Then he eats 13 berries. How many berries does Steven have now?

○ 5

○ 27

○ 37

○ 40

GO ON ➡

Name _____

**Beginning-of-Year Test
Page 10**

37. What is the difference?

$$80 - 26$$

○ 4 tens 4 ones

○ 4 tens 6 ones

○ 5 tens 4 ones

○ 8 tens 2 ones

38. Which statement is **true?**

○ 1 centimeter is longer than 1 meter.

○ 1 meter is longer than 1 centimeter.

○ 1 meter is shorter than 1 centimeter.

○ 1 meter is the same as 1 centimeter.

39. Measure the length of the rope to the nearest centimeter. Which length is the **best** choice?

○ 5 centimeters   ○ 8 centimeters

○ 11 centimeters   ○ 13 centimeters

40. Use a centimeter ruler. Measure the length of each object.

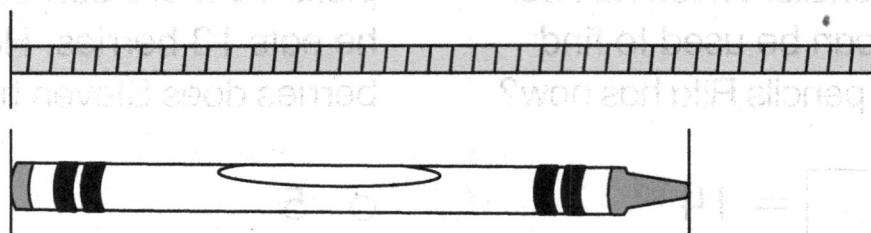

How much longer is the ribbon than the crayon?

○ 3 centimeters longer   ○ 9 centimeters longer

○ 12 centimeters longer   ○ 21 centimeters longer

Assessment Guide
© Houghton Mifflin Harcourt Publishing Company

AG10

Beginning-of-Year Test

Name _____

**Beginning-of-Year Test
Page 11**

41. Allen asked some friends to name their favorite color.
    2 children like red.
    5 children like blue.
    3 children like green.
    1 child likes yellow.

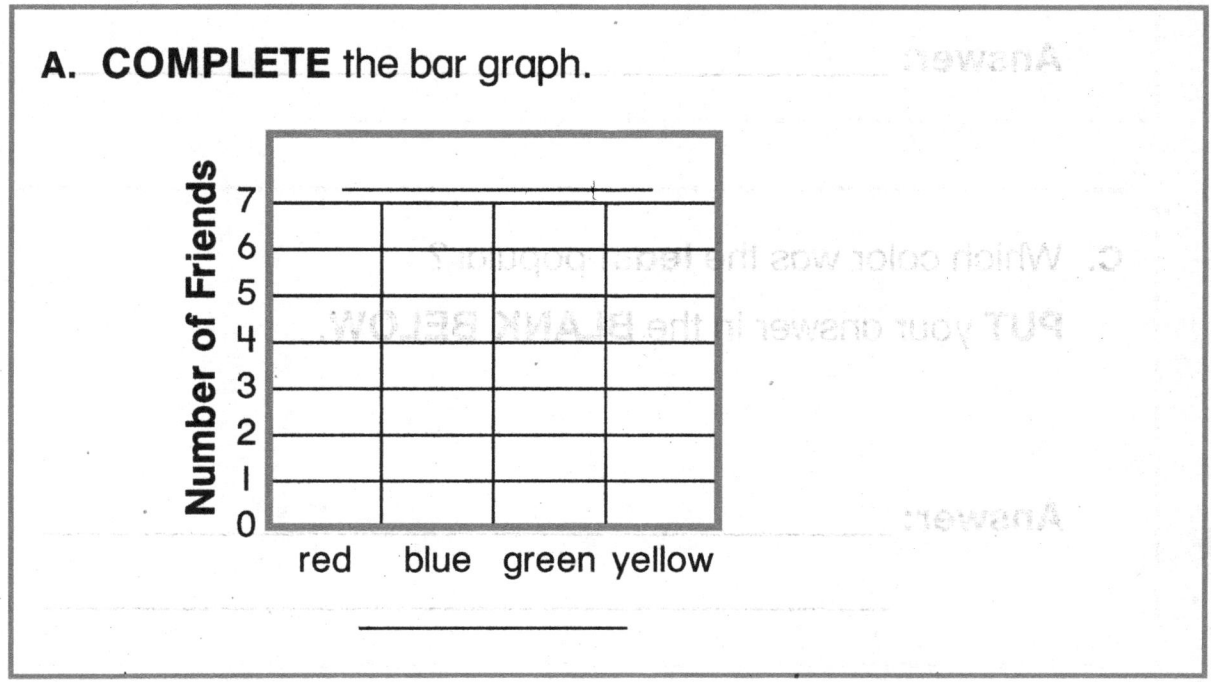

A. **COMPLETE** the bar graph.

Name _____

Beginning-of-Year Test
Page 12

B. Which color was the **most** popular?

PUT your answer in the **BLANK BELOW**.

Answer: _____

C. Which color was the **least** popular?

PUT your answer in the **BLANK BELOW**.

Answer: _____
_____

Assessment Guide
© Houghton Mifflin Harcourt Publishing Company

AG12

Beginning-of-Year Test

Name _____

**Middle-of-Year Test
Page 1**

Choose the correct answer.

1. Which shows a related subtraction fact?

    $9 + 5 = 14$

    ○ $19 - 14 = 5$
    ○ $14 + 5 = 19$
    ○ $14 - 5 = 9$
    ○ $9 - 5 = 4$

2. There were 16 birds at the park. Then 9 birds flew away. Which number sentence shows how many birds are at the park now?

    ○ $16 + 9 = 25$
    ○ $16 - 9 = 7$
    ○ $9 - 7 = 2$
    ○ $2 + 7 = 9$

3. Fran picks 3 red flowers, 7 yellow flowers, and 3 pink flowers. How many flowers does Fran pick in all?

    ○ 6
    ○ 10
    ○ 12
    ○ 13

4. There are 429 students at the museum. There are 180 boys. How many girls are at the museum?

Hundreds	Tens	Ones
☐	☐	☐
4	2	9
− 1	8	0

    ○ 239   ○ 249
    ○ 349   ○ 369

Assessment Guide
© Houghton Mifflin Harcourt Publishing Company

AG13

Middle-of-Year Test

Name _____

Middle-of-Year Test
Page 2

5. Rhonda collected 256 shells to make necklaces. Jan collected 315 shells. How many shells did the two girls collect?

○ 476   ○ 561
○ 571   ○ 601

6. What is the difference?

   507
 − 368

○ 139   ○ 149
○ 239   ○ 241

7. Use an inch ruler. What is the length of the pencil to the nearest inch?

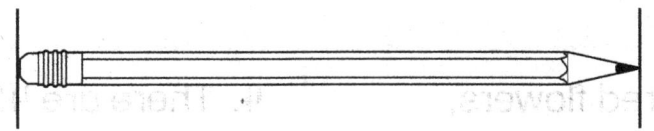

○ 2 inches   ○ 3 inches
○ 4 inches   ○ 5 inches

8. Larry made a line plot to show the lengths of his toy cars. How many cars are shown in the line plot?

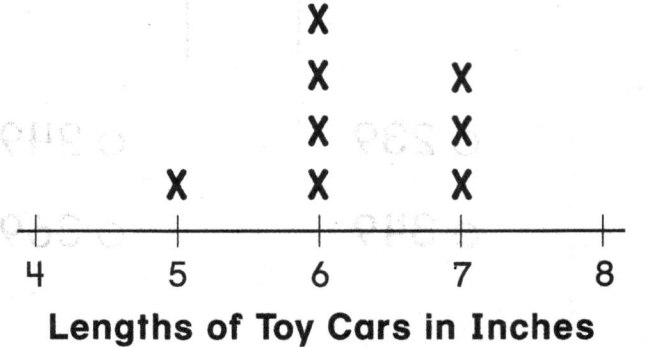

Lengths of Toy Cars in Inches

○ 7
○ 8
○ 9
○ 10

GO ON ➡

Assessment Guide
© Houghton Mifflin Harcourt Publishing Company

AG14

Middle-of-Year Test

Name _____

Middle-of-Year Test
Page 3

9. Which is the **best** estimate of the length of a real park bench?

- ○ 1 foot
- ○ 6 feet
- ○ 15 feet
- ○ 20 feet

10. Frank wants to measure the length of a bus. Which is the **best** tool for Frank to use?

- ○ yardstick
- ○ counters
- ○ cup
- ○ pencil

11. Ms. Ikeda writes an even number on the board. Which could be the number that Ms. Ikeda writes?

- ○ 7
- ○ 11
- ○ 13
- ○ 14

12. What is the value of the underlined digit?

4<u>5</u>

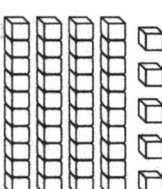

- ○ 4
- ○ 40
- ○ 5
- ○ 50

GO ON ➡

Assessment Guide
© Houghton Mifflin Harcourt Publishing Company

AG15

Middle-of-Year Test

Name _____

Middle-of-Year Test
Page 4

13. Which shows another way to write the number?

    32

    ○ 3 tens 2 ones

    ○ twenty-three

    ○ 20 + 3

    ○ 3 + 2

14. Vanessa starts at 460 and counts by tens. What are the next 6 numbers Vanessa will say?

    ○ 470, 480, 490, 500, 510, 520

    ○ 465, 470, 475, 480, 485, 490

    ○ 462, 464, 466, 468, 470, 472

    ○ 461, 462, 463, 464, 465, 466

15. Which object is shaped like a cone?

○   ○

○   ○

16. Danny makes a triangle with straws. He uses one straw for each side of the shape. How many straws does Danny need?

    ○ 3

    ○ 4

    ○ 5

    ○ 6

GO ON

Assessment Guide
© Houghton Mifflin Harcourt Publishing Company

AG16

Middle-of-Year Test

Name _____

Middle-of-Year Test
Page 5

17. Which of these shapes has **fewer** than 5 angles?

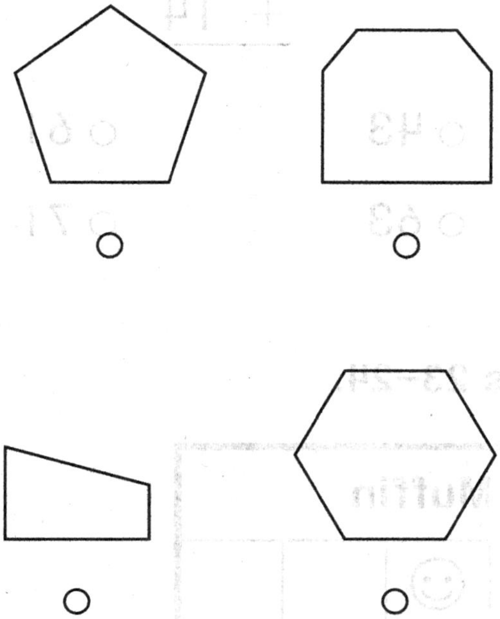

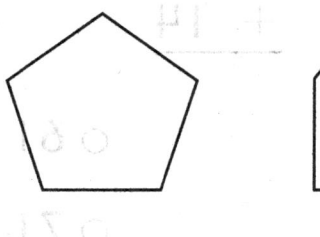

○   ○

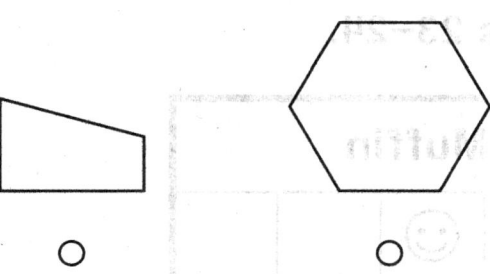

○   ○

18. Which shows a third of the shape shaded?

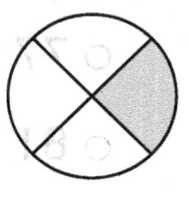   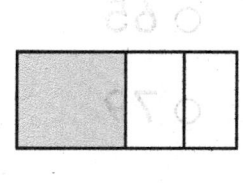

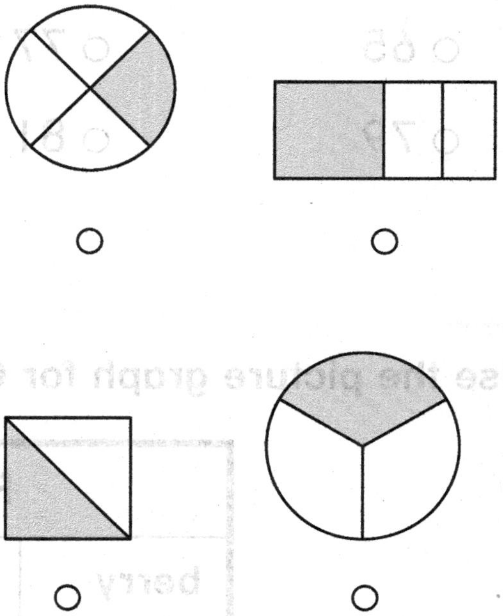

○   ○

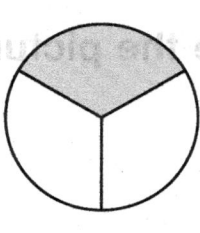

○   ○

19. Beth has 26 stickers. Ken has 51 stickers. Which number sentence tells how many stickers Beth and Ken have?

○ 51 − 26 = 25

○ 26 + 51 = 77

○ 26 + 25 = 51

○ 15 + 26 = 41

20. What is the sum?

$$\begin{array}{r} 18 \\ 32 \\ +12 \\ \hline \end{array}$$

○ 44

○ 50

○ 52

○ 62

GO ON ➔

Assessment Guide
© Houghton Mifflin Harcourt Publishing Company

AG17

Middle-of-Year Test

Name _____

21. Jacob has 46 marbles. Chloe has 31 marbles. How many marbles do they have?

○ 65   ○ 77
○ 79   ○ 81

22. What is the sum?

$$\begin{array}{r} 57 \\ +\ 14 \\ \hline \end{array}$$

○ 43   ○ 61
○ 63   ○ 71

Use the picture graph for Questions 23–24.

Favorite Muffin					
berry	☺	☺	☺		
corn	☺	☺	☺	☺	☺
pumpkin	☺				

Key: Each ☺ stands for 1 child.

23. How many **more** children chose berry than pumpkin?

○ 1   ○ 2
○ 3   ○ 5

24. **2 more** children choose pumpkin. How many ☺ should be in the pumpkin row now?

○ 3   ○ 4
○ 5   ○ 7

Name _____

Middle-of-Year Test
Page 7

25. Julio made a tally chart of the vegetables he planted.

Vegetables Planted	
Vegetable	Tally
beans	ⅢⅠ Ⅰ
carrots	ⅠⅠⅠⅠ
peas	ⅢⅠ ⅠⅠ

How many peas did Julio plant?

○ 4   ○ 6
○ 7   ○ 10

26. Use the bar graph.

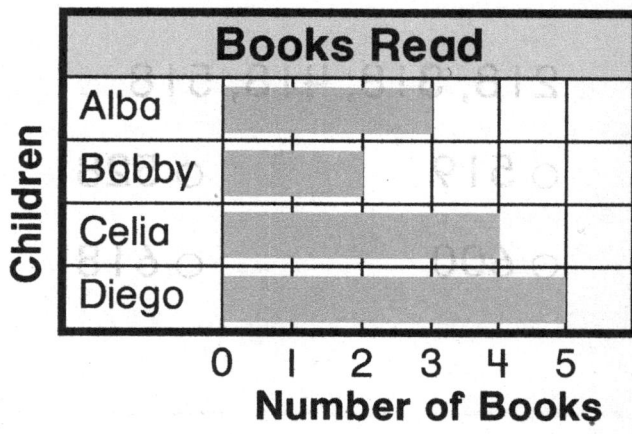

Who read the **most** books?

○ Alba   ○ Celia
○ Bobby   ○ Diego

27. Which number has the digit 8 in the tens place?

○ 298
○ 586
○ 803
○ 8,000

28. Which shows another way to write the number?

seven hundred thirteen

○ 7 + 13
○ 70 + 13
○ 70 + 10 + 3
○ 700 + 10 + 3

GO ON

AG19

Name _____

Middle-of-Year Test
Page 8

29. Look at the pattern. What number comes next?

218, 318, 418, 518, ▓

○ 519   ○ 528
○ 600   ○ 618

30. Which comparison is true?

○ 253 > 315
○ 315 > 381
○ 354 < 366
○ 472 < 425

31. Melinda gave her friend these coins. What is the value of the coins?

45 cents   55 cents   60 cents   75 cents
   ○   ○   ○   ○

32. Clara wants to buy a comb that costs one dollar. Which coins have a total value of one dollar?

○ 20 dimes

○ 20 nickels

○ 5 dimes

○ 5 nickels

GO ON

Assessment Guide
© Houghton Mifflin Harcourt Publishing Company

AG20

Middle-of-Year Test

Name _____

Middle-of-Year Test
Page 9

33. Jared ate lunch at the time shown on the clock. At what time did Jared eat lunch?

12:15 A.M.   3:00 A.M.   12:15 P.M.   3:00 P.M.
   ○   ○   ○   ○

34. Break apart the ones to subtract. What is the difference?

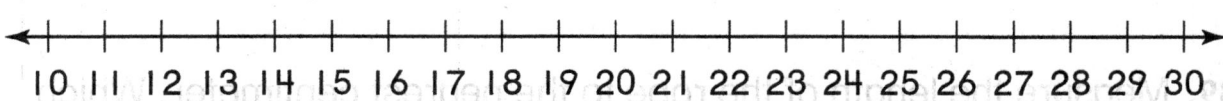

$$24 - 6 = \underline{}$$

17   18   19   20
○   ○   ○   ○

35. Ana has 36 cups. She fills 12 cups with juice. Which number sentence can be used to find how many cups are empty?

○ 12 + ☐ = 30
○ 24 − ☐ = 12
○ 36 + 12 = ☐
○ 36 − 12 = ☐

36. Mark has 59 shells. He gives 21 shells to Ethan and 16 shells to Beth. How many shells does Mark have now?

○ 22
○ 37
○ 38
○ 43

GO ON ➡

Assessment Guide   AG21   Middle-of-Year Test
© Houghton Mifflin Harcourt Publishing Company

Name _____

Middle-of-Year Test
Page 10

37. What is the difference?

$$\begin{array}{r} 90 \\ -\ 38 \\ \hline \end{array}$$

○ 6 tens 8 ones

○ 6 tens 2 ones

○ 5 tens 8 ones

○ 5 tens 2 ones

38. Which statement is **true**?

○ I meter is the same as I centimeter.

○ I meter is shorter than I centimeter.

○ I centimeter is longer than I meter.

○ I centimeter is shorter than I meter.

39. Measure the length of the rope to the nearest centimeter. Which length is the best choice?

○ 6 centimeters   ○ 8 centimeters

○ 10 centimeters   ○ 12 centimeters

40. Use a centimeter ruler. Measure the length of each object.

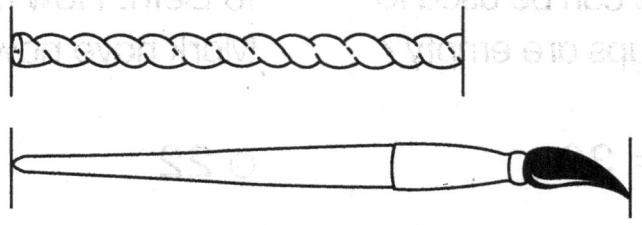

How much longer is the paintbrush than the string?

○ 2 centimeters longer   ○ 6 centimeters longer

○ 10 centimeters longer   ○ 16 centimeters longer

Assessment Guide
© Houghton Mifflin Harcourt Publishing Company

AG22

Middle-of-Year Test

Name _____

Middle-of-Year Test
Page 11

41. Patricia asked some friends to name their favorite animal.

4 children like cats.

5 children like horses.

6 children like dogs.

2 children likes gerbils.

A. COMPLETE the bar graph.

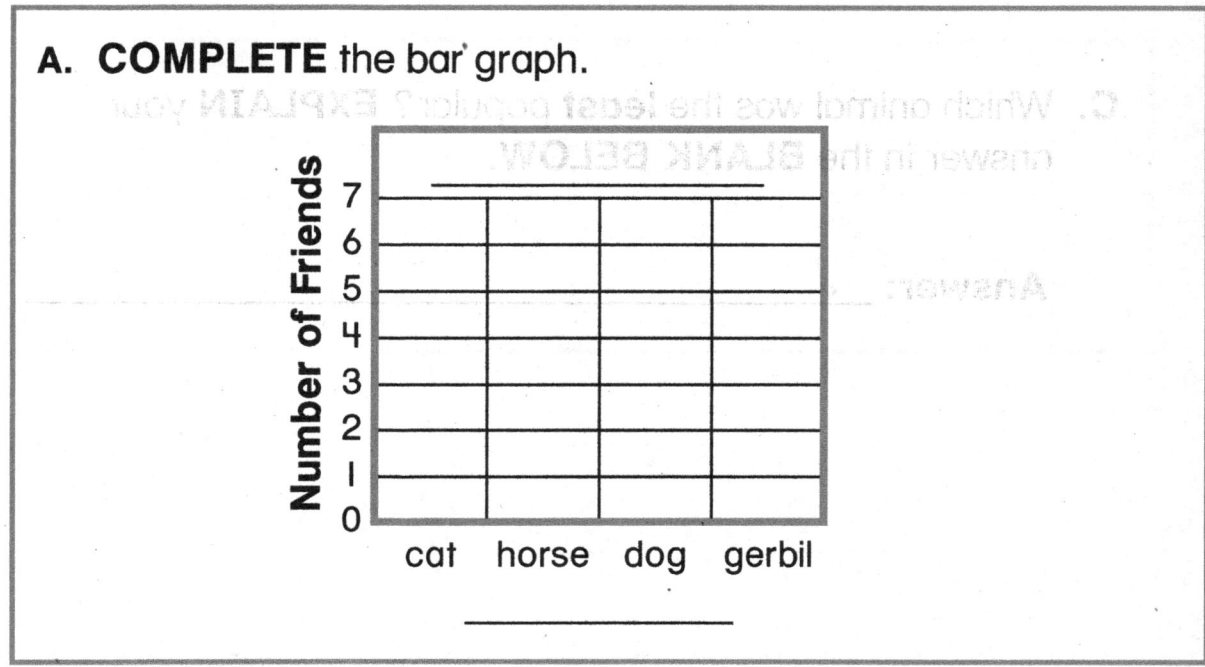

Name _____

Middle-of-Year Test
Page 12

B. Which animal was the **most** popular? **EXPLAIN** your answer in the **BLANK BELOW**.

Answer: _____

C. Which animal was the **least** popular? **EXPLAIN** your answer in the **BLANK BELOW**.

Answer: _____

Name _____

End-of-Year Test
Page 1

Choose the correct answer.

1. Which shows a related addition fact?

    $17 - 9 = 8$

    ○ $17 + 9 = 26$
    ○ $9 - 8 = 1$
    ○ $8 + 9 = 17$
    ○ $25 - 8 = 17$

2. There are 7 big dogs and 6 small dogs. Which number sentence shows how many dogs there are in all?

    ○ $7 + 6 = 13$
    ○ $7 - 1 = 6$
    ○ $10 + 7 = 17$
    ○ $19 - 7 = 12$

3. Tess collects 2 green leaves, 8 red leaves, and 5 yellow leaves. How many leaves does Tess collect in all?

    ○ 7
    ○ 10
    ○ 13
    ○ 15

4. There are 545 seats at the theater. 362 seats are filled. How many seats are empty?

Hundreds	Tens	Ones
☐	☐	☐
5	4	5
− 3	6	2

    ○ 123   ○ 183
    ○ 223   ○ 283

GO ON

Assessment Guide
© Houghton Mifflin Harcourt Publishing Company

AG25

End-of-Year Test

Name _____

End-of-Year Test
Page 2

5. Josh collected 233 baseball cards for his collection. Dave collected 428 cards. How many cards did the two boys collect?

○ 396   ○ 435
○ 551   ○ 661

6. What is the difference?

$$\begin{array}{r} 803 \\ -\ 427 \\ \hline \end{array}$$

○ 376   ○ 386
○ 476   ○ 486

7. Use an inch ruler. What is the length of the paintbrush to the nearest inch?

○ 4 inches   ○ 5 inches
○ 6 inches   ○ 7 inches

8. Billy made a line plot to show the lengths of his toy trains. How many trains are shown in the line plot?

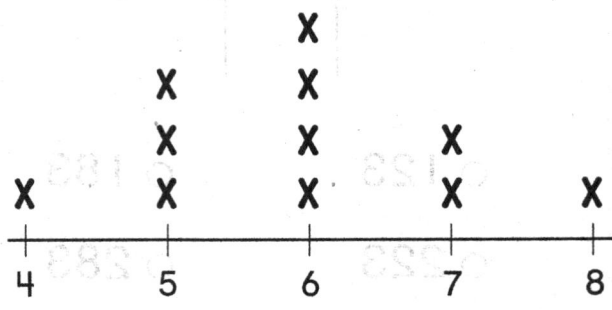

Lengths of Toy Trains in Inches

○ 9
○ 10
○ 11
○ 12

GO ON

Assessment Guide
© Houghton Mifflin Harcourt Publishing Company

AG26

End-of-Year Test

Name _____

End-of-Year Test
Page 3

9. Which is the **best** estimate of the length of an adult's shoe?

- ○ 1 foot
- ○ 3 feet
- ○ 5 feet
- ○ 9 feet

10. Eddie wants to measure the distance around a water bottle. Which is the **best** tool for Eddie to use?

- ○ cup
- ○ measuring tape
- ○ pencil
- ○ counters

11. Ms. Martinez writes an even number on the board. Which could be the number that Ms. Martinez writes?

- ○ 9
- ○ 10
- ○ 11
- ○ 13

12. What is the value of the underlined digit?

<u>6</u>2

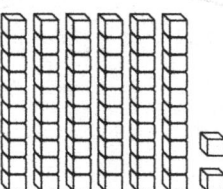

- ○ 2
- ○ 6
- ○ 20
- ○ 60

GO ON

Assessment Guide
© Houghton Mifflin Harcourt Publishing Company

AG27

End-of-Year Test

End-of-Year Test
Page 4

13. Which shows another way to write the number?

    74

    ○ 4 tens 7 ones

    ○ forty-seven

    ○ 70 + 4

    ○ 7 + 4

14. Kelly starts at 180 and counts by tens. What are the next 6 numbers Kelly will say?

    ○ 181, 182, 183, 184, 185, 186

    ○ 182, 184, 186, 188, 190, 192

    ○ 185, 190, 195, 200, 205, 210

    ○ 190, 200, 210, 220, 230, 240

15. Which object is shaped like a cube?

    ○   ○

       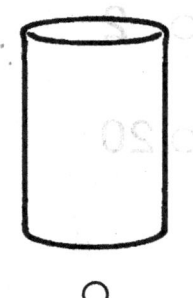
    ○   ○

16. Danny makes a hexagon with straws. He uses one straw for each side of the shape. How many straws does Danny need?

    ○ 3

    ○ 4

    ○ 5

    ○ 6

GO ON

Assessment Guide AG28 End-of-Year Test
© Houghton Mifflin Harcourt Publishing Company

Name _____

End-of-Year Test
Page 5

17. Which of these shapes has **more** than 5 sides?

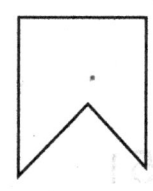

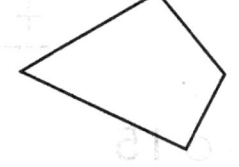

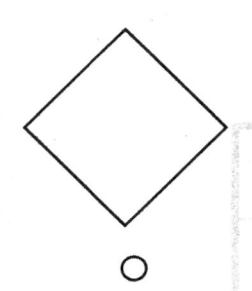

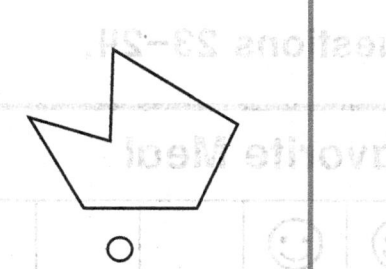

○   ○

○   ○

18. Which shows a fourth of the shape shaded?

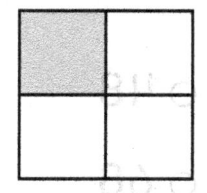

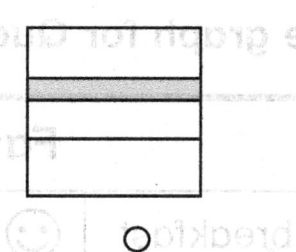

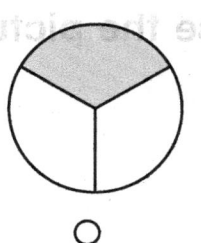

○   ○

○   ○

19. Jen has 52 beads. She buys 17 more beads. Which number sentence tells how many beads Jen has now?

○ 52 + 17 = 69
○ 52 − 17 = 35
○ 35 + 17 = 52
○ 25 + 17 = 42

20. What is the sum?

  43
  27
+ 13
────

○ 40
○ 70
○ 73
○ 83

GO ON ➡

Assessment Guide
© Houghton Mifflin Harcourt Publishing Company

AG29

End-of-Year Test

Name _____

End-of-Year Test
Page 6

21. Anne has 34 crayons. Carey has 14 crayons. How many crayons do they have?

○ 20   ○ 48
○ 50   ○ 68

22. What is the sum?

$$\begin{array}{r} 38 \\ +\ 23 \\ \hline \end{array}$$

○ 15   ○ 51
○ 55   ○ 61

Use the picture graph for Questions 23–24.

Favorite Meal					
breakfast	☺	☺			
lunch	☺	☺	☺		
dinner	☺	☺	☺	☺	☺

Key: Each ☺ stands for 1 child.

23. How many **more** children chose dinner than breakfast?

○ 1   ○ 2
○ 3   ○ 5

24. **2 more** children choose lunch. How many ☺ should be in the lunch row now?

○ 2   ○ 4
○ 5   ○ 7

GO ON

Assessment Guide
© Houghton Mifflin Harcourt Publishing Company

AG30

End-of-Year Test

Name _____

End-of-Year Test
Page 7

25. Mr. Campa made a tally chart of the trees he sold.

Trees Sold	
Tree	Tally
apple	IIII IIII
oak	III
pine	IIII I

How many pine trees did Mr. Campa sell?

○ 3   ○ 6
○ 9   ○ 10

26. Use the bar graph.

How We Get to School

Way: Bike, Bus, Car, Walk

Number of Children: 0 1 2 3 4 5

How many children do **not** take the bus to school?

○ 1   ○ 2
○ 3   ○ 6

27. Which number has the digit 2 in the hundreds place?

○ 25
○ 298
○ 742
○ 2,000

28. Which shows another way to write the number?

four hundred twenty-three

○ 4 + 2 + 3
○ 40 + 23
○ 400 + 20 + 3
○ 400 + 20 + 30

GO ON →

Assessment Guide
© Houghton Mifflin Harcourt Publishing Company

AG31

End-of-Year Test

29. Look at the pattern. What number comes next?

351, 361, 371, 381,

○ 382   ○ 386

○ 391   ○ 481

30. Which comparison is **true**?

○ 323 > 304

○ 295 > 405

○ 215 < 154

○ 165 < 123

31. Steve gave his sister these coins. What is the value of the coins?

65 cents   75 cents   80 cents   90 cents
   ○   ○   ○   ○

32. Hugo wants to buy a bottle of juice that costs one dollar. Which coins have a total value of one dollar?

○ 10 quarters

○ 10 nickels

○ 4 quarters

○ 4 nickels

Name _____

End-of-Year Test
Page 9

33. Emily went to bed for the night at the time shown on the clock. At what time did Emily go to bed?

5:40 A.M.   8:25 A.M.   5:40 P.M.   8:25 P.M.
   ○   ○   ○   ○

---

34. Break apart the ones to subtract. What is the difference?

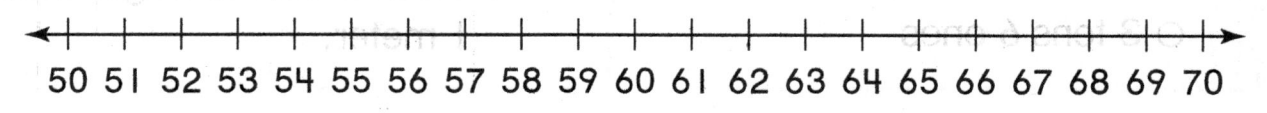

50 51 52 53 54 55 56 57 58 59 60 61 62 63 64 65 66 67 68 69 70

$$62 - 6 = \_\_\_$$

56   58   60   68
○   ○   ○   ○

---

35. Mia had 48 stickers. She gave away 17 stickers. Which number sentence can be used to find how many stickers Mia has now?

○ 40 + ☐ = 48

○ 48 − 17 = ☐

○ 48 + 17 = ☐

○ 65 − ☐ = 48

36. Jamal has a box with 35 crayons. He puts 22 more crayons in the box. Then he takes 14 crayons out of the box. How many crayons are in the box now?

○ 8

○ 43

○ 53

○ 57

GO ON

Assessment Guide
© Houghton Mifflin Harcourt Publishing Company

AG33

End-of-Year Test

Name _____

End-of-Year Test
Page 10

37. What is the difference?

$$70 - 44$$

- ○ 2 tens 4 ones
- ○ 2 tens 6 ones
- ○ 3 tens 4 ones
- ○ 3 tens 6 ones

38. Which statement is **true?**

- ○ I meter is longer than I centimeter.
- ○ I meter is shorter than I centimeter.
- ○ I centimeter is the same as I meter.
- ○ I centimeter is longer than I meter.

39. Measure the length of the rope to the nearest centimeter. Which length is the **best** choice?

- ○ 6 centimeters
- ○ 9 centimeters
- ○ 8 centimeters
- ○ 11 centimeters

40. Use a centimeter ruler. Measure the length of each object.

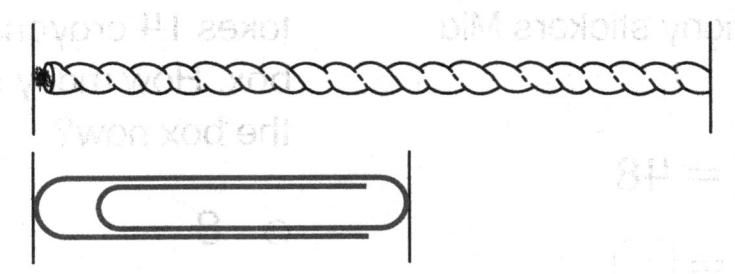

How much longer is the string than the paperclip?

- ○ 4 centimeters longer
- ○ 9 centimeters longer
- ○ 5 centimeters longer
- ○ 14 centimeters longer

GO ON ➡

Assessment Guide
© Houghton Mifflin Harcourt Publishing Company

AG34

End-of-Year Test

Name ____________________

41. Brooke asked some friends to name their favorite food to eat at a baseball game.

3 children like peanuts.

2 children like popcorn.

4 children like hamburgers.

5 children like hot dogs.

A. COMPLETE the bar graph.

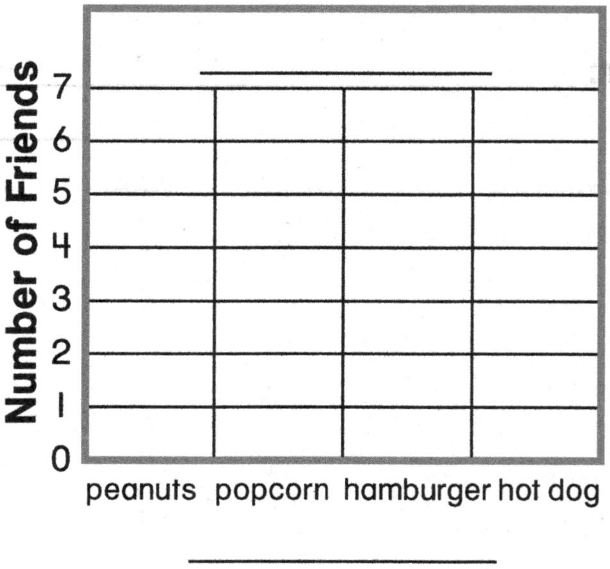

GO ON

Name _____

End-of-Year Test
Page 12

B. Which food was the **most** popular? **EXPLAIN** your answer in the **BLANKS BELOW**.

Answer: _____

C. Which food was the **least** popular? **EXPLAIN** your answer in the **BLANKS BELOW**.

Answer: _____
_____

Name _____

Chapter 1 Test
Page 1

Choose the correct answer.

1. Which shows $7 + 7 = 14$?

   ○   ○

   ○   ○

2. Which sum is an even number?

   ○ $4 + 5 = 9$

   ○ $4 + 4 = 8$

   ○ $3 + 4 = 7$

   ○ $4 + 1 = 5$

3. Paul counts by twos to 20.

   Karen counts by ones to 10.

   Maria counts by fives to 50.

   Who will say **more** numbers?

   ○ Paul

   ○ Karen

   ○ Maria

   ○ They all say the same amount.

4. Which of the numbers shows counting by fives?

   ○ 1, 5, 9, 13, 17

   ○ 5, 7, 9, 11, 13

   ○ 10, 20, 30, 40, 50

   ○ 15, 20, 25, 30, 35

GO ON

Assessment Guide
© Houghton Mifflin Harcourt Publishing Company

AG37

Chapter 1 Test

Name _____

Chapter 1 Test
Page 2

5. Which is another way to describe 27?

- ○ 20 + 7
- ○ 20 + 70
- ○ 2 + 7
- ○ 70 + 2

6. Which is another way to describe 52?

- ○ 5 + 2
- ○ 20 + 5
- ○ 50 + 2
- ○ 500 + 2

7. Which shows how many tens and ones in 65?

- ○ 5 tens 6 ones
- ○ 6 tens 0 ones
- ○ 6 tens 5 ones
- ○ 6 tens 6 ones

8. Which is another way to write thirty-eight?

- ○ 8
- ○ 38
- ○ 83
- ○ 308

GO ON

Assessment Guide
© Houghton Mifflin Harcourt Publishing Company

AG38

Chapter 1 Test

Name _____

Chapter 1 Test
Page 3

9. Which is another way to write sixty-four?

　○ 64 + 4

　○ 4 tens 6 ones

　○ 60 + 40

　○ 64

10. Which is another way to write the number?

10 + 6

　○ 16

　○ 26

　○ 61

　○ 66

11. Count by tens.

435, 445, 455, 465, _____

What number comes next?

　○ 466

　○ 470

　○ 475

　○ 565

12. Which group of numbers shows counting by hundreds?

　○ 100, 105, 110, 115

　○ 200, 201, 202, 203

　○ 300, 400, 500, 600

　○ 400, 410, 420, 430

GO ON

Assessment Guide
© Houghton Mifflin Harcourt Publishing Company

AG39

Chapter 1 Test

Name _____

Chapter 1 Test
Page 4

13. Ms. Brice buys 37 markers for the classroom. What choice is missing from the list?

Packs of 10 markers	Single markers
3	7
2	17
0	37

○ 3 packs, 7 markers

○ 1 pack, 27 markers

○ 1 pack, 17 markers

○ 0 packs, 27 markers

14. Ann needs 12 folders for school. What choice is missing from the list?

Packs of 10 folders	Single folders
0	12

○ 2 packs, 0 folders

○ 2 packs, 1 folder

○ 1 pack, 12 folders

○ 1 pack, 2 folders

15. Jon wants to buy 21 apples. What choice is missing from the list?

Bags of 10 apples	Single apples
2	1
1	11

○ 0 bags, 21 apples

○ 0 bags, 11 apples

○ 1 bag, 21 apples

○ 2 bags, 2 apples

16. The blocks show the number 30. Which is a way to show this number?

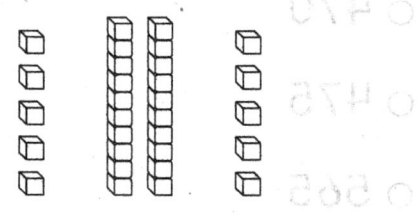

○ 1 ten 5 ones

○ 1 ten 10 ones

○ 2 tens 5 ones

○ 2 tens 10 ones

GO ON

Assessment Guide
© Houghton Mifflin Harcourt Publishing Company

AG40

Chapter 1 Test

Name _____

Chapter 1 Test
Page 5

17. Cameron is thinking of a number that has a digit that is **less than** 4 in the tens place. It has a digit **greater than** 6 in the ones place. What could Cameron's number be?

○ 88

○ 60 + 9

○ thirty-seven

○ five tens 4 ones

18. Which is a way to show the number 56?

○ 4 tens 6 ones

○ 4 tens 16 ones

○ 5 tens 16 ones

○ 5 tens 26 ones

19. Which ten frame shows an even number?

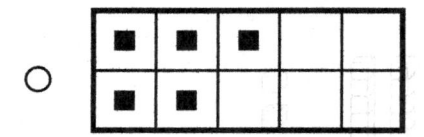

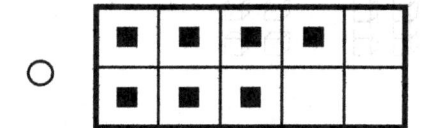

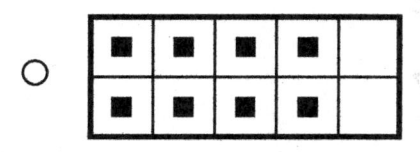

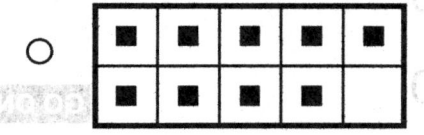

○
○
○
○

20. The Morris family has an even number of dogs and an odd number of cats. Which could be the number of pets in the Morris family?

○ 1 dog and 2 cats

○ 1 dog and 3 cats

○ 2 dogs and 1 cat

○ 2 dogs and 2 cats

GO ON

Assessment Guide
© Houghton Mifflin Harcourt Publishing Company

AG41

Chapter 1 Test

Name _____

Chapter 1 Test
Page 6

21. Katherine has an even number of bracelets and an odd number of necklaces. Which group of bracelets and necklaces could belong to Katherine?

○ 4 bracelets and 5 necklaces

○ 3 bracelets and 2 necklaces

○ 4 bracelets and 6 necklaces

○ 3 bracelets and 5 necklaces

22. Ally wrote this riddle:

My number has 9 tens and 1 one.

Which number matches Ally's clues?

○ 9

○ 11

○ 19

○ 91

23. What is the value of the digit 3 in 53?

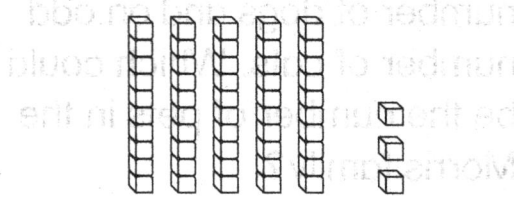

○ 1
○ 3
○ 5
○ 30

24. What is the value of the underlined digit?

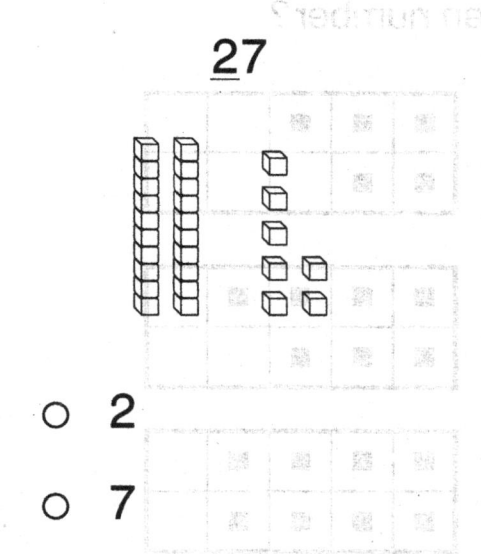

27

○ 2
○ 7
○ 20
○ 70

GO ON

Assessment Guide
© Houghton Mifflin Harcourt Publishing Company

AG42

Chapter 1 Test

Name _____

Chapter 1 Test
Page 7

25. Mr. Arnold needs 24 pencils. He can buy them in packs of 10 pencils or as single pencils.

A. COMPLETE THE CHART to find all the different ways Mr. Arnold can buy the pencils.

Packs of 10 pencils	Single pencils

B. CHOOSE two of the ways and **EXPLAIN** how these ways show the same number of pencils.

PUT your answer in the **BLANK BELOW**.

Answer: _____
_____
_____

Assessment Guide
© Houghton Mifflin Harcourt Publishing Company

Name _____

**Chapter 1 Test
Page 8**

C. COMPLETE THE CHART to show 4 different ways that Mr. Arnold could buy 24 pencils if he could buy them in packs of 8, packs of 10, or as singles.

Packs of 10 pencils	Packs of 8 pencils	Single pencils

D. Which of the ways from your chart in Part C use the fewest single pencils?

WRITE your answer on the **BLANK BELOW**.

Answer: _____

Name _____

**Chapter 2 Test
Page 1**

Choose the correct answer.

1. Look at the picture. Which has the same value as 12 tens?

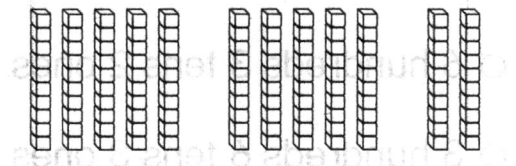

   ○ 2 tens
   ○ 2 hundreds
   ○ 1 hundred 1 ten
   ○ 1 hundred 2 tens

2. Sonya has 140 beads. How many bags of 10 beads does she need so that she will have 200 beads in all?

   ○ 6
   ○ 14
   ○ 20
   ○ 60

3. What is **10 more than** 529?

   ○ 429
   ○ 519
   ○ 539
   ○ 629

4. A store has 263 board games. It has **100 fewer** puzzles than board games. The store has **10 more** action figures than puzzles. How many action figures does it have?

   ○ 163
   ○ 173
   ○ 273
   ○ 353

Assessment Guide
© Houghton Mifflin Harcourt Publishing Company

AG45

GO ON

Chapter 2 Test

Name _____

**Chapter 2 Test
Page 2**

5. Count the hundreds, tens, and ones. Which shows the same number?

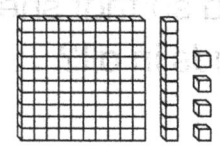

   ○ 400 + 40 + 1
   ○ 400 + 10 + 4
   ○ 100 + 40 + 1
   ○ 100 + 10 + 4

6. Ray sold 362 tickets to the show. Which is another way to write the number?

   ○ 6 hundreds 3 tens 2 ones
   ○ 3 hundreds 6 tens 3 ones
   ○ 3 hundreds 6 tens 2 ones
   ○ 2 hundreds 6 tens 3 ones

7. Sue's box has 264 pens. Which of these numbers is greater than 264?

   ○ 188
   ○ 246
   ○ 258
   ○ 310

8. Dana has 562 paper clips. Mitchell has fewer paper clips than Dana. Which number tells how many paper clips Mitchell could have?

   ○ 526
   ○ 563
   ○ 572
   ○ 661

GO ON

Assessment Guide
© Houghton Mifflin Harcourt Publishing Company

Name _____

Chapter 2 Test
Page 3

9. Which number has the digit 6 in the hundreds place?

   ○ 68
   ○ 196
   ○ 362
   ○ 610

10. There are 203 birds in a zoo. What is the value of the digit 3 in the number 203?

    ○ 3
    ○ 30
    ○ 200
    ○ 300

11. Claudia's collection has four hundred sixty-five stickers. Which is another way to write the number?

    ○ 400 + 60 + 5
    ○ 400 + 600 + 5
    ○ 40 + 60 + 5
    ○ 4 + 6 + 5

12. Ollie draws a quick picture.

    Which is another way to write Ollie's number?

    ○ 1 hundred 4 tens 8 ones
    ○ 100 + 8 + 4
    ○ 1 hundred 8 tens 4 ones
    ○ 100 + 80 + 7

GO ON

Assessment Guide
© Houghton Mifflin Harcourt Publishing Company

AG47

Chapter 2 Test

Name _____

Chapter 2 Test
Page 4

13. How many hundreds does the picture show?

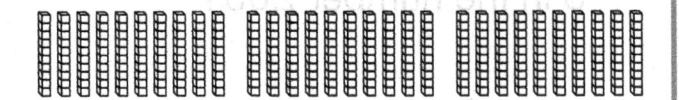

    ○ 30 hundreds
    ○ 20 hundreds
    ○ 3 hundreds
    ○ 2 hundreds

14. Which number has the same value as 80 tens?

    ○ 8
    ○ 80
    ○ 800
    ○ 801

15. What number is shown with these blocks?

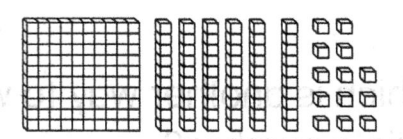

    ○ 133
    ○ 136
    ○ 163
    ○ 173

16. Rafi made this model of 309.

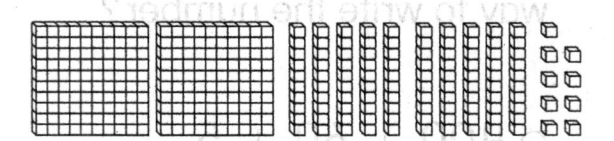

    Which shows how many hundreds, tens, and ones?

    ○
Hundreds	Tens	Ones
3	19	0

    ○
Hundreds	Tens	Ones
3	10	9

    ○
Hundreds	Tens	Ones
2	0	19

    ○
Hundreds	Tens	Ones
2	10	9

GO ON

Assessment Guide
© Houghton Mifflin Harcourt Publishing Company

AG48

Chapter 2 Test

Name _____

Chapter 2 Test
Page 5

17. What is the next number in the counting pattern?

472, 572, 672, 772

○ 773
○ 782
○ 872
○ 873

18. Rico wrote this counting pattern. What two numbers are next in Rico's pattern?

183, 283, 383, 483

○ 484, 485
○ 493, 503
○ 583, 683
○ 683, 783

19. Terry has 164 marbles. Which is another way to write the number 164?

○ one hundred sixteen
○ one hundred forty-six
○ one hundred sixty-four
○ four hundred sixty-one

20. There are two hundred four students in Marcy's school. Which shows this number?

○ 24
○ 204
○ 214
○ 240

GO ON

Assessment Guide
© Houghton Mifflin Harcourt Publishing Company

AG49

Chapter 2 Test

Name _____

Chapter 2 Test
Page 6

21. Compare the numbers. Which symbol makes the comparison **true**?

241 ◯ 214

○ >
○ <
○ =
○ +

22. Which of the following comparisons is **true**?

○ 120 < 115
○ 343 < 328
○ 691 > 706
○ 705 > 609

23. What number is shown with these blocks?

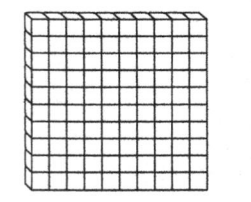

○ 12
○ 102
○ 120
○ 201

24. A store sells 2 boxes of 100 pencils and some single pencils. Which number shows how many pencils the store could sell?

○ 120
○ 182
○ 200
○ 206

GO ON

Assessment Guide
© Houghton Mifflin Harcourt Publishing Company

AG50

Chapter 2 Test

Name _____

Chapter 2 Test
Page 7

25. Jill and Ed collect postcards. Jill has 124 postcards. Ed has 131 postcards.

> **A.** Who has more postcards?
>
> **PUT** your answer in the **BLANK BELOW**.
>
>
>
> **Answer:** _____

> **B.** Jill gets 10 more postcards. Ed gets 5 more postcards. Who has more postcards now?
>
> **SHOW all your WORK**.
>
>
>
>
>
> **Answer:** _____

Name _____

**Chapter 2 Test
Page 8**

C. The next week, Jill gets 18 more postcards. Ed gets 15 more postcards. Who has more postcards now?

SHOW all your WORK.

Answer: _____

Name _____

Chapter 3 Test
Page 1

Choose the correct answer.

1. What is the sum?

    $5 + 2$

    ○ 4
    ○ 5
    ○ 6
    ○ 7

2. What is the difference?

    $13 - 9$

    ○ 4
    ○ 5
    ○ 6
    ○ 7

3. Which number sentence has the same difference as $15 - 8 = \blacksquare$?

    ○ $10 - 1 = \blacksquare$
    ○ $10 - 2 = \blacksquare$
    ○ $10 - 3 = \blacksquare$
    ○ $10 - 4 = \blacksquare$

4. Which doubles fact could you use to find the sum of $4 + 5$?

    ○ $1 + 1$
    ○ $2 + 2$
    ○ $3 + 3$
    ○ $4 + 4$

GO ON ➡

Assessment Guide
© Houghton Mifflin Harcourt Publishing Company

AG53

Chapter 3 Test

Name _____

Chapter 3 Test
Page 2

5. Owen puts 4 apples in each basket. There are 6 baskets. How many apples are there in all?

   ○ 16
   ○ 20
   ○ 24
   ○ 28

6. Which shows a way to find the sum?

   $4 + 8 =$ _____

   ○ $10 + 2 = 12$
   ○ $10 + 4 = 14$
   ○ $10 + 6 = 16$
   ○ $10 + 8 = 18$

7. Ava grows 3 red flowers, 4 yellow flowers, and 4 purple flowers in her garden. How many flowers does Ava grow in all?

   ○ 7
   ○ 8
   ○ 10
   ○ 11

8. What is the sum?

   $4 + 5 + 7 =$ _____

   ○ 9
   ○ 11
   ○ 16
   ○ 17

GO ON

Assessment Guide
© Houghton Mifflin Harcourt Publishing Company

AG54

Chapter 3 Test

Name _____

Chapter 3 Test
Page 3

9. John has 16 rocks. He gives 7 rocks to his cousin. How many rocks does John have now?

   ○ 7
   ○ 8
   ○ 9
   ○ 10

10. What is the missing number in the related subtraction fact?

    $9 + 3 = 12$

    $12 - 9 = \boxed{}$

    ○ 3
    ○ 4
    ○ 6
    ○ 9

11. Eli has 13 marbles. Amber has 6 marbles. How many **more** marbles does Eli have than Amber?

    ○ 7
    ○ 8
    ○ 10
    ○ 19

12. Look at the picture. Which addition sentence shows the total number of shapes?

    ▲ ▲ ▲ ▲ ▲
    ▲ ▲ ▲ ▲ ▲
    ▲ ▲ ▲ ▲ ▲

    ○ $3 + 5 = 8$
    ○ $3 + 3 + 3 = 9$
    ○ $5 + 5 + 5 = 15$
    ○ $5 + 5 + 5 + 5 + 5 = 25$

GO ON ➡

Assessment Guide
© Houghton Mifflin Harcourt Publishing Company

AG55

Chapter 3 Test

Name _____

Chapter 3 Test
Page 4

13. David has 7 pencils in his pencil case. He has 1 pencil in his desk. How many pencils does David have?

○ 6
○ 7
○ 8
○ 9

14. Bob has 15 markers and 7 crayons. How many more markers than crayons does Bob have?

○ 7
○ 8
○ 12
○ 15

15. There are 11 books on a shelf. Then Reba takes some books off the shelf. Now there are 4 books on the shelf. How many books did Reba take off the shelf?

○ 7
○ 8
○ 10
○ 15

16. Peter sees 8 dogs. Beth sees 1 more dog than Peter. How many dogs do they see in all?

○ 14
○ 16
○ 17
○ 18

GO ON

Assessment Guide
© Houghton Mifflin Harcourt Publishing Company

AG56

Chapter 3 Test

Name _____

Chapter 3 Test
Page 5

17. Mia puts 5 crackers on each plate. How many crackers does she put on 3 plates?

   ○ 10
   ○ 15
   ○ 20
   ○ 25

18. There are 8 large plates and 7 small plates on a table. How many plates are on the table?

   ○ 10
   ○ 12
   ○ 15
   ○ 16

19. Kate sees 4 white dogs, 9 brown dogs, and 6 black dogs at the park. How many dogs does she see at the park?

   ○ 10
   ○ 15
   ○ 16
   ○ 19

20. Mark picks 11 apples. Anna picks 5 apples. Which number sentence shows how many **fewer** apples Anna picked than Mark?

   ○ 16 − 11 = 5
   ○ 11 − 5 = 6
   ○ 11 + 5 = 16
   ○ 11 + 6 = 17

GO ON ➡

Assessment Guide
© Houghton Mifflin Harcourt Publishing Company

21. Which shows a related addition fact?

$$13 - 6 = 7$$

○ $6 + 7 = 13$
○ $7 + 13 = 20$
○ $7 - 6 = 1$
○ $13 + 6 = 19$

22. Julian had 14 grapes. He gave 5 grapes to Lindsay. How many grapes does Julian have now?

○ 9
○ 10
○ 11
○ 19

23. Leah has 8 green apples and 4 red apples. How many apples does Leah have?

○ 4
○ 8
○ 12
○ 14

24. Mae has 2 rows of stickers. There are 4 stickers in each row. How many stickers does Mae have?

○ 2
○ 4
○ 6
○ 8

Name _____

Chapter 3 Test
Page 7

25. There are 3 oranges in a bag. Mr. Johnson puts **4 more** oranges in the bag.

   A. How many oranges are in the bag now?

   PUT your answer in the **BLANK BELOW**.

   Answer: _____ oranges

   B. Mr. Johnson puts 3 more oranges in the bag. How many oranges are in the bag now?

   PUT your answer in the **BLANK BELOW**.

   Answer: _____ oranges

Name _____

C. Mr. Johnson added 5 more oranges. How many oranges are there in the bag now?

PUT your answer in the **BLANK BELOW**.

Answer: _____ oranges

Name _____

Chapter 4 Test
Page 1

Choose the correct answer.

1. Which shows a way to find the sum?

    41 + 29

    ○ 41 + 30 = 71
    ○ 40 + 20 = 60
    ○ 40 + 30 = 70
    ○ 50 + 30 = 80

2. Which shows a way to find the sum?

    66 + 16

    ○ 60 + 16 = 76
    ○ 70 + 16 = 86
    ○ 60 + 12 = 72
    ○ 70 + 12 = 82

3. What is the sum?

    58
    24
    + 3
    ___

    ○ 27
    ○ 82
    ○ 85
    ○ 95

4. What is the sum?

    54
    31
    + 17
    ___

    ○ 48
    ○ 71
    ○ 85
    ○ 102

GO ON

Assessment Guide
© Houghton Mifflin Harcourt Publishing Company

AG61

Chapter 4 Test

Name _____

Chapter 4 Test
Page 2

5. Jake has 57 stamps. He buys 37 more stamps. How many stamps does Jake have now?

- ○ 20
- ○ 80
- ○ 87
- ○ 94

6. There are 45 red grapes in a bowl. There are 18 green grapes in the bowl. How many grapes are in the bowl?

- ○ 37
- ○ 53
- ○ 63
- ○ 68

7. Lisa has 17 markers in her desk. She puts 29 more markers in the desk. How many markers are in the desk now?

- ○ 12
- ○ 32
- ○ 36
- ○ 46

8. There are 33 papers on a table. Ms. Smith puts 25 more papers on the table. How many papers are on the table now?

- ○ 12
- ○ 18
- ○ 58
- ○ 68

GO ON

Assessment Guide
© Houghton Mifflin Harcourt Publishing Company

AG62

Chapter 4 Test

Name _____

Chapter 4 Test
Page 3

9. Jenna counts 35 birds on a roof. She counts 26 birds on the ground. How many birds does Jenna see?

　○ 51
　○ 61
　○ 65
　○ 70

10. What is the sum?

$$\begin{array}{r} 23 \\ +\ 18 \\ \hline \end{array}$$

　○ 30
　○ 31
　○ 40
　○ 41

11. What is the sum?

$$\begin{array}{r} 58 \\ 17 \\ 23 \\ +\ 31 \\ \hline \end{array}$$

　○ 98
　○ 119
　○ 125
　○ 129

12. On four days this week, Caroline practiced her flute for 41 minutes, 38 minutes, 33 minutes, and 14 minutes. What is the number of minutes that Caroline practiced her flute?

　○ 98
　○ 107
　○ 113
　○ 126

GO ON

Assessment Guide
© Houghton Mifflin Harcourt Publishing Company

AG63

Chapter 4 Test

Name _____

Chapter 4 Test
Page 4

13. Gina and Eric play a game. Gina scores 26 points. Eric scores 31 points. How many points do Gina and Eric score?

   ○ 55
   ○ 57
   ○ 67
   ○ 77

14. What is the sum?

$$\begin{array}{r} 56 \\ +\ 67 \\ \hline \end{array}$$

   ○ 93
   ○ 107
   ○ 113
   ○ 123

15. What is the sum?

   25 + 18 = ▢

   ○ 33
   ○ 43
   ○ 53
   ○ 268

16. The table shows the number of books four children read this summer. How many books did Dora and Jason read?

| Books Read This Summer ||
Child	Number of Books
Dora	31
Kyle	38
Mark	29
Jason	27

   ○ 50   ○ 58
   ○ 69   ○ 321

GO ON

Assessment Guide
© Houghton Mifflin Harcourt Publishing Company

AG64

Chapter 4 Test

Name _____

Chapter 4 Test
Page 5

17. Savannah sees 18 butterflies at the park. Then she sees **8 more**. How can you find how many butterflies Savannah sees?

○ 18 − 8
○ 18 + 6
○ 18 − 5 − 2
○ 18 + 2 + 6

18. Break apart ones to make a ten. What is the sum?

$$89 + 5$$

○ 84
○ 92
○ 94
○ 104

19. James and Flora have 38 markers in all. Flora has 16 markers. How many markers does James have?

○ 22
○ 44
○ 54
○ 62

20. Alex has 46 beads. Chris has 33 beads. How many beads do they have?

○ 59
○ 63
○ 76
○ 79

GO ON

Assessment Guide
© Houghton Mifflin Harcourt Publishing Company

Name _____

**Chapter 4 Test
Page 6**

21. What is the sum?

$$\begin{array}{r} 16 \\ +\ 18 \\ \hline \end{array}$$

○ 23
○ 24
○ 33
○ 34

22. What is the sum?

$$\begin{array}{r} 59 \\ +\ 27 \\ \hline \end{array}$$

○ 75
○ 76
○ 85
○ 86

23. Jonas bought 13 blue marbles. He bought 38 purple marbles. How many marbles did Jonas buy?

○ 41
○ 45
○ 49
○ 51

24. Mark reads 13 pages of a book on Thursday. He reads 47 on Friday. How many pages does he read?

○ 34
○ 50
○ 60
○ 65

GO ON ➤

Assessment Guide
© Houghton Mifflin Harcourt Publishing Company

AG66

Chapter 4 Test

25. Renee spends 22 minutes doing science homework. She spends 14 minutes doing history homework. She spends 33 minutes doing math homework.

A. How many minutes did Renee spend doing homework?

PUT your answer in the BLANK BELOW.

Answer: _____ minutes

B. Renee spends another 32 minutes doing English homework. How many minutes did Renee spend doing homework now?

PUT your answer in the BLANK BELOW.

Answer: _____ minutes

GO ON

Name _____

Chapter 4 Test
Page 8

C. Renee spends another 25 minutes doing math homework. How many minutes did Renee spend doing homework now?

PUT your answer in the **BLANK BELOW**.

Answer: _____

Name _____

Chapter 5 Test
Page 1

Choose the correct answer.

1. There are 52 berries in a basket. Jan takes 16 berries out of the basket. How many berries are in the basket now?

    ○ 34
    ○ 36
    ○ 44
    ○ 69

2. Which shows a different way to write the subtraction problem?

    $$48 - 23$$

    ○ 40
      −20
      ―――
       20

    ○ 18
      − 3
      ―――
       15

    ○ 48
      −23
      ―――
       25

    ○ 25
      −23
      ―――
        2

3. Break apart the number you are subtracting. What is the difference?

    $$44 - 9 = \underline{}$$

    35   40   45   53
    ○   ○   ○   ○

4. Use the number line. Count up to find the difference.

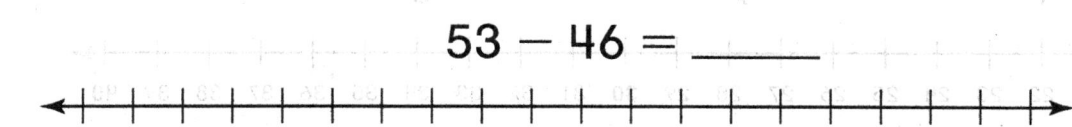

    $$53 - 46 = \underline{}$$

    ←―+―+―+―+―+―+―+―+―+―+―+―+―+―+―+―+―+―+―+―→
      41 42 43 44 45 46 47 48 49 **50** 51 52 53 54 55 56 57 58 59 **60**

    What is the difference?

    6   7   8   9
    ○   ○   ○   ○

    GO ON ▶

Assessment Guide   AG69   Chapter 5 Test
© Houghton Mifflin Harcourt Publishing Company

Name _____

Chapter 5 Test
Page 2

5. What is the difference?

$$\begin{array}{r} 6\,2 \\ -\,2\,5 \\ \hline \end{array}$$

○ 37   ○ 43
○ 47   ○ 87

6. In which problem do you need to regroup to subtract?

○ 57 − 26
○ 39 − 18
○ 30 − 19
○ 73 − 63

7. Regroup if you need to. What is the difference?

Tens	Ones
4	6
− 1	9

○ 23   ○ 27
○ 33   ○ 37

8. Break apart the number you are subtracting. What is the difference?

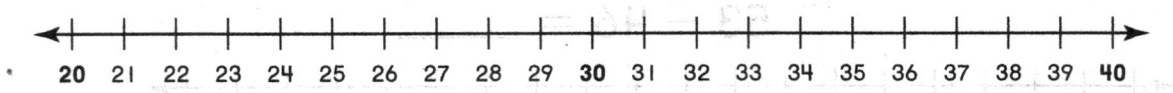

20 21 22 23 24 25 26 27 28 29 30 31 32 33 34 35 36 37 38 39 40

38 − 16 = _____

2   12   22   32
○   ○   ○   ○

Assessment Guide   AG70   Chapter 5 Test
© Houghton Mifflin Harcourt Publishing Company

GO ON

Name _____

Chapter 5 Test
Page 3

9. Mrs. Dobbs has 47 stickers. She gives away 12 stickers. How many stickers does she have left?

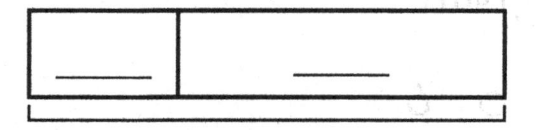

○ 5
○ 18
○ 35
○ 59

10. Hannah makes 21 bracelets for the fair. Ladonne makes 5 more bracelets than Hannah. How many bracelets do they make in all?

○ 16
○ 26
○ 47
○ 52

11. Abby has 25 stamps. She buys 16 more stamps. Then she gives 12 stamps to Jeff. How many stamps does Abby have now?

○ 13
○ 29
○ 41
○ 53

12. There were 53 books on the shelf. Some of the books were checked out. Now there are 25 books on the shelf. Which number sentence can be used to find how many books were checked out?

○ 53 − ■ = 25
○ 53 + 25 = ■
○ 28 + 53 = ■
○ 28 − ■ = 25

GO ON →

Assessment Guide   AG71   Chapter 5 Test
© Houghton Mifflin Harcourt Publishing Company

Name _____

Chapter 5 Test
Page 4

13. Subtract 14 from 43. Which shows the difference?

Tens	Ones

○ 2 tens 9 ones

○ 3 tens 1 one

○ 5 tens 7 ones

○ 8 tens 7 ones

14. Jason has 52 baseball cards. Kim has 28 baseball cards. How many more baseball cards does Jason have than Kim?

○ 6

○ 24

○ 34

○ 36

15. Break apart ones to subtract. What is the difference?

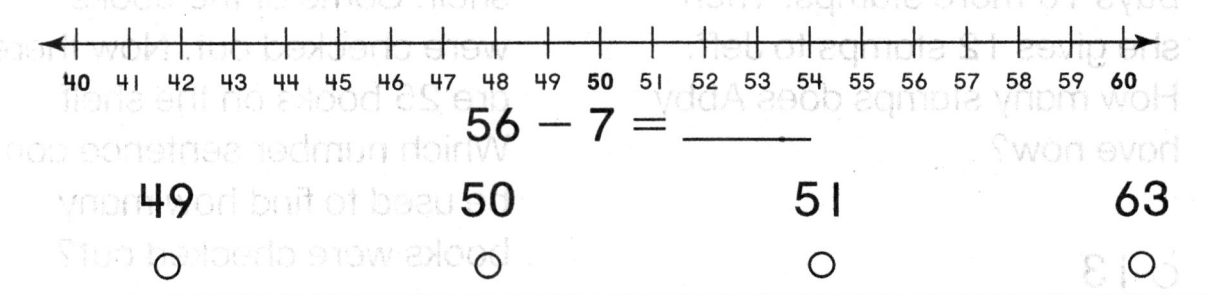

$56 - 7 =$ _____

49   50   51   63
○   ○   ○   ○

16. Use the number line. Count up to find the difference.

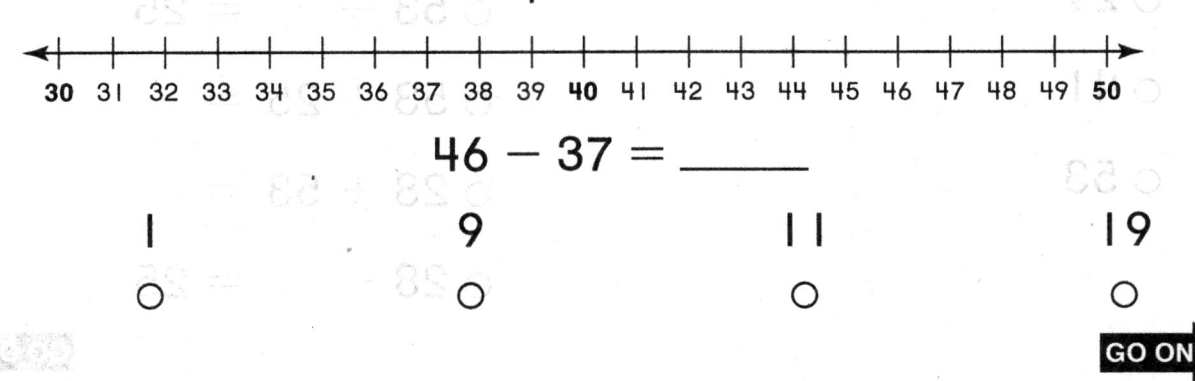

$46 - 37 =$ _____

1   9   11   19
○   ○   ○   ○

GO ON

Assessment Guide
© Houghton Mifflin Harcourt Publishing Company

Name _____

**Chapter 5 Test
Page 5**

17. What is the difference?

$$\begin{array}{r} 50 \\ -26 \\ \hline \end{array}$$

○ 24
○ 25
○ 36
○ 76

18. What is the difference?

$$\begin{array}{r} 35 \\ -14 \\ \hline \end{array}$$

○ 11
○ 21
○ 29
○ 49

19. What is the difference?

Tens	Ones
☐	☐
6	2
− 2	9

○ 33
○ 47
○ 81
○ 91

20. Draw a quick picture to solve. What is the difference?

Tens	Ones
☐	☐
4	2
− 1	7

Tens	Ones

○ 25
○ 35
○ 39
○ 59

GO ON ➡

Assessment Guide
© Houghton Mifflin Harcourt Publishing Company

AG73

Chapter 5 Test

Name _____

Chapter 5 Test
Page 6

21. Break apart the number you are subtracting. What is the difference?

$$51 - \underline{17} = \underline{}$$

34   44   46   68
○   ○   ○   ○

---

22. Juan has 16 candles. Melanie has **3 more** candles than Juan. How many candles do they have altogether?

○ 13   ○ 19

○ 32   ○ 35

23. Rosa has some stickers. She gives away 35 stickers. Now Rosa has 19 stickers. Which number sentence can be used to find how many stickers Rosa had to start?

○ 35 − ■ = 19

○ ■ − 35 = 19

○ ■ + 19 = 35

○ 54 + ■ = 73

---

24. A store sells 34 shirts. 7 shirts are blue. The rest of the shirts are white. How many white shirts does the store sell?

17   21   27   41
○   ○   ○   ○

GO ON

Name _____

Chapter 5 Test
Page 7

25. Dan has 29 animal pictures. Kayla has 37 animal pictures.

> **A.** Who has **more** animal pictures? How many **more**?
> Use a bar model to solve.
>
> **SHOW** your **WORK**.
>
>
>
> **Answer:** _____ has _____ more animal pictures.

Assessment Guide
© Houghton Mifflin Harcourt Publishing Company

AG75

GO ON

Chapter 5 Test

Name _____

Chapter 5 Test
Page 8

B. Dan gets **11 more** animal pictures. Who has more animal pictures now? How many more? Use a bar model to solve.
SHOW your **WORK**.

Answer: _____ has _____ more animal pictures.

C. Dan is given 9 **more** pictures and is told to share them with Kayla so that they have the same amount of pictures. How many pictures should he give her? How many will they both have then?
WRITE your answer in the blank below.

Answer: _____
_____

Name _____

**Chapter 6 Test
Page 1**

Choose the correct answer.

1. What is the difference?

Hundreds	Tens	Ones
	☐	☐
3	8	2
− 1	5	4

○ 128   ○ 232
○ 228   ○ 238

2. Meg had 432 stickers. She gave 216 of the stickers to a classmate. How many stickers does Meg have left?

○ 216
○ 224
○ 226
○ 248

3. What is the difference?

Hundreds	Tens	Ones
☐	☐	☐
3	1	7
− 1	2	5

○ 182
○ 192
○ 292
○ 432

4. What is the sum?

Hundreds	Tens	Ones
☐	☐	
4	7	1
+ 2	3	4

○ 245
○ 605
○ 705
○ 715

GO ON ▶

Assessment Guide
© Houghton Mifflin Harcourt Publishing Company

AG77

Chapter 6 Test

Name _____

Chapter 6 Test
Page 2

5. Add 213 and 151. What is the sum?

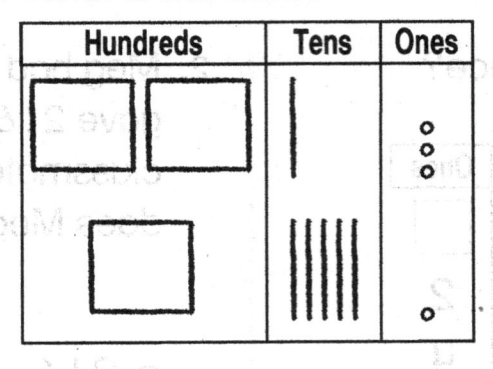

Hundreds	Tens	Ones
☐ ☐ ☐	| |||||	ooo o

○ 263   ○ 264   ○ 314   ○ 364

---

6. A bird watcher counted 163 white birds and 185 black birds. How many birds did she count?

$$\begin{array}{r} 163 \\ +185 \\ \hline \end{array} \longrightarrow \begin{array}{r} 100+60+3 \\ +100+80+5 \\ \hline \end{array}$$

○ 248   ○ 258   ○ 348   ○ 358

---

7. Sydney's class collected 368 cans for recycling. Ramon's class collected 413 cans. How many cans did the two classes collect in all?

○ 771   ○ 781
○ 871   ○ 881

8. What is the sum?

Hundreds	Tens	Ones
☐		
5	2	7
+ 1	4	8

○ 575   ○ 665
○ 675   ○ 765

GO ON

Assessment Guide
© Houghton Mifflin Harcourt Publishing Company

AG78

Chapter 6 Test

Name _____

Chapter 6 Test
Page 3

9. What is the sum?

```
   2 9 9
 + 2 3 7
```

○ 426

○ 436

○ 526

○ 536

10. At the store, there are 863 rocks and shells. There are 121 rocks. How many shells are there?

○ 642

○ 742

○ 842

○ 984

11. What is the difference?

```
   4 6 2
 - 1 9 5
```

○ 267

○ 333

○ 367

○ 377

12. What is the difference?

```
   5 0 6
 - 3 2 9
```

○ 177

○ 183

○ 187

○ 223

Assessment Guide
© Houghton Mifflin Harcourt Publishing Company

AG79

GO ON

Chapter 6 Test

Name _____

Chapter 6 Test
Page 4

13. What is the difference?

Hundreds	Tens	Ones
	☐	☐
4	5	2
− 2	3	7

○ 115   ○ 125
○ 215   ○ 225

14. What is the sum?

Hundreds	Tens	Ones
☐	☐	
5	8	2
+ 2	3	7

○ 719   ○ 755
○ 819   ○ 829

15. What is the sum?

Hundreds	Tens	Ones
☐	☐	
6	7	4
+ 1	8	1

○ 493
○ 755
○ 765
○ 855

16. What is the sum?

Hundreds	Tens	Ones
	☐	
3	2	8
+ 5	1	4

○ 832
○ 834
○ 842
○ 942

GO ON

Assessment Guide
© Houghton Mifflin Harcourt Publishing Company

AG80

Chapter 6 Test

Name _____

Chapter 6 Test
Page 5

17. Add 254 and 215. What is the sum?

Hundreds	Tens	Ones
☐ ☐ ☐ ☐	|||||	ooooo ooooo

○ 454   ○ 459   ○ 464   ○ 469

18. Which shows 524 broken apart into hundreds, tens, and ones?

○ 500 + 20 + 40
○ 500 + 20 + 4
○ 50 + 20 + 4
○ 50 + 2 + 4

19. What is the sum?

$$\begin{array}{r} 1\ 5\ 8 \\ +\ 1\ 6\ 2 \\ \hline \end{array}$$

○ 210   ○ 220
○ 320   ○ 330

20. What is the sum?

$$\begin{array}{r} 2\ 3\ 8 \\ +\ 2\ 2\ 5 \\ \hline \end{array}$$

○ 413   ○ 453
○ 463   ○ 563

GO ON

Assessment Guide
© Houghton Mifflin Harcourt Publishing Company

AG81

Chapter 6 Test

Name _____

Chapter 6 Test
Page 6

21. There are 417 children at the festival. 288 of the children are girls. Which problem shows how many of the children are boys?

- ○ 395
   − 218

- ○ 480
   − 341

- ○ 417
   − 288

- ○ 360
   − 324

22. There are 594 children at the museum. There are 235 boys. How many girls are at the museum?

- ○ 359
- ○ 361
- ○ 369
- ○ 869

23. A farmer has 305 sheep. She moves 263 sheep into a field. Which sentence describes a step in finding how many sheep are left?

- ○ Regroup 1 ten as 10 ones.
- ○ Subtract 0 tens from 6 tens.
- ○ Regroup 1 hundred as 10 tens.
- ○ Subtract 2 hundreds from 3 hundreds.

24. What is the difference?

   701
 − 546

- ○ 145
- ○ 155
- ○ 245
- ○ 265

GO ON

Assessment Guide
© Houghton Mifflin Harcourt Publishing Company

AG82

Chapter 6 Test

Name _____

Chapter 6 Test
Page 7

25. Riley used this method to find the sum of 538 + 254.

```
   538
 + 254
   700
    80
 +  12
   792
```

A. **DESCRIBE** how Riley solves addition problems.

PUT your answer in the **BLANK BELOW**.

Answer: _____

_____

_____

Name _____

**Chapter 6 Test
Page 8**

B. USE Riley's method to **FIND** the following sum.

```
  364
+ 217
```

Answer: _____

C. Claire tried to use Riley's method to add 129 + 415. **FIND** her mistake.

PUT your answer on the **BLANKS BELOW**.

```
   129
 + 415
 -----
   500
    30
 +   4
 -----
   534
```

Answer: _____
_____

Assessment Guide
© Houghton Mifflin Harcourt Publishing Company

AG84

Chapter 6 Test

Name _____

Chapter 7 Test
Page 1

Choose the correct answer.

1. What is the time on the clock?

    ○ 6:00
    ○ 8:00
    ○ 8:30
    ○ 9:30

2. What is the time on the clock?

    ○ 9:00
    ○ 9:30
    ○ 12:00
    ○ 12:30

3. What is the time on the clock?

    ○ 7:40
    ○ 7:50
    ○ 8:00
    ○ 8:50

4. What is the time on the clock?

    ○ 1:03
    ○ 1:15
    ○ 3:05
    ○ 3:10

GO ON

Assessment Guide
© Houghton Mifflin Harcourt Publishing Company

AG85

Chapter 7 Test

Name _____

Chapter 7 Test
Page 2

5. What is the total value of this money?

○ $1.15   ○ $1.26

○ $1.30   ○ $1.35

6. Kim has 60 cents. Which set of coins shows this amount?

7. Which combination of coins has a value of 75 cents?

○ 2 quarters

○ 10 nickels and 5 dimes

○ 1 quarter, 2 dimes, and 3 nickels

○ 6 dimes, 2 nickels, and 5 pennies

8. Count on. What is the total value of these coins?

○ 17¢   ○ 28¢

○ 32¢   ○ 37¢

GO ON

Assessment Guide
© Houghton Mifflin Harcourt Publishing Company

AG86

Chapter 7 Test

Name _____

Chapter 7 Test
Page 3

9. Which clock shows quarter past 8?

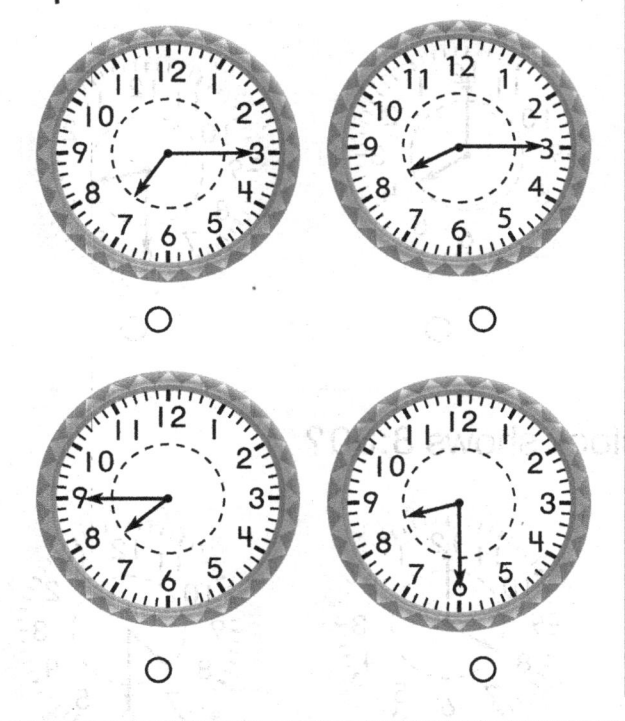

○   ○

○   ○

10. Kareem has these coins in his wallet.

What is the total value of these coins?

○ 31¢   ○ 36¢

○ 51¢   ○ 77¢

11. Jessie buys a book that costs $1.00. Which coins does she use to buy the book?

○ 5 pennies

○ 5 dimes

○ 10 nickels

○ 10 dimes

12. One night, Mrs. Rivera saw a rocket launch at the time shown on the clock.

At what time did Mrs. Rivera see the rocket launch?

○ 12:40 A.M.   ○ 12:40 P.M.

○ 8:00 A.M.   ○ 8:00 P.M.

GO ON

Name _____

Chapter 7 Test
Page 4

13. Petra's soccer practice starts at 4:00. Which clock shows 4:00?

○   ○   ○   ○

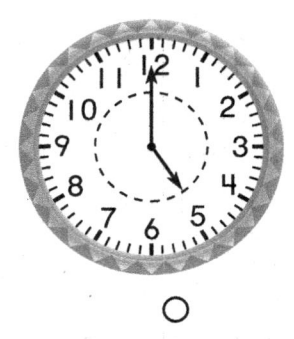

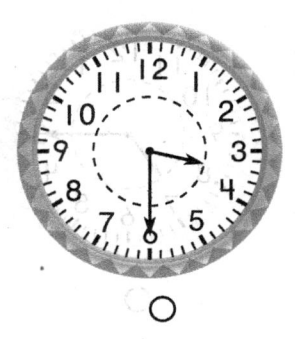

14. Lee goes to bed at 8:30. Which clock shows 8:30?

○   ○   ○   ○

15. Sasha eats breakfast at the time shown on the clock.

At what time does Sasha eat breakfast?

○ 7:08   ○ 7:40

○ 8:08   ○ 8:40

16. Count on. What is the total value of these coins?

○ 7¢

○ 11¢

○ 25¢

○ 45¢

GO ON

Assessment Guide
© Houghton Mifflin Harcourt Publishing Company

AG88

Chapter 7 Test

Name _____

Chapter 7 Test
Page 5

17. Fred has these coins in his pocket.

How much money does Fred have in his pocket?

○ 15¢   ○ 25¢
○ 55¢   ○ 75¢

18. Count on. What is the total value of these coins?

○ 8¢
○ 13¢
○ 25¢
○ 28¢

19. Which clock shows 20 minutes past 3?

○   ○   ○   ○

20. What is the total value of these coins?

33¢   42¢   51¢   60¢
○   ○   ○   ○

GO ON ➡

Assessment Guide
© Houghton Mifflin Harcourt Publishing Company

AG89

Chapter 7 Test

Chapter 7 Test
Page 6

Name _____

21. Which group of coins has a total value of $1.00?

   ○ (3 quarters, 2 dimes, 1 nickel)

   ○ (2 quarters, 3 dimes, 1 dime)

   ○ (2 quarters, 1 dime, 1 dime, 1 nickel, 1 nickel)

22. James paid for a drink with this money. How much did James pay?

   ○ $1.41   ○ $1.46   ○ $1.56   ○ $1.61

23. Marta pays $1.80 for a snack. What is one way to show $1.80?

   ○ one $1 bill, 2 quarters, and 3 nickels
   ○ one $1 bill, 3 quarters, and 1 dime
   ○ 5 quarters, 2 dimes, and 1 nickel
   ○ 7 quarters and 1 nickel

24. The clock shows the time that Kyle got on the school bus this morning. At what time did he get on the bus?

   ○ 7:30 P.M.   ○ 7:30 A.M.
   ○ 9:35 P.M.   ○ 9:35 A.M.

GO ON

Assessment Guide

Name _____

Chapter 7 Test
Page 7

25. Hannah gave her sister these coins.

> A. **WRITE** the value of the coins. **EXPLAIN** how you found the total value. **PUT** your answer on the lines below.
>
> Answer: _____
> _____

> B. **DRAW** more coins so the total is 96¢.
>
> Answer: _____

Name _____

Chapter 7 Test
Page 8

C. SUBTRACT 1 penny, 1 nickel, and 1 dime from 96¢. How many cents are left over? **WRITE** your answer on the line below.

Answer: _____

D. SUBTRACT 1 penny, 2 nickels, and 1 quarter from 96¢. Then add 2 dimes. How many cents do you have now? **SHOW** all your **WORK**.

Answer: _____
_____

Assessment Guide
© Houghton Mifflin Harcourt Publishing Company

AG92

Chapter 7 Test

Name _____

Chapter 8 Test
Page 1

Choose the correct answer.

Use the line plot for Exercises 1-2.

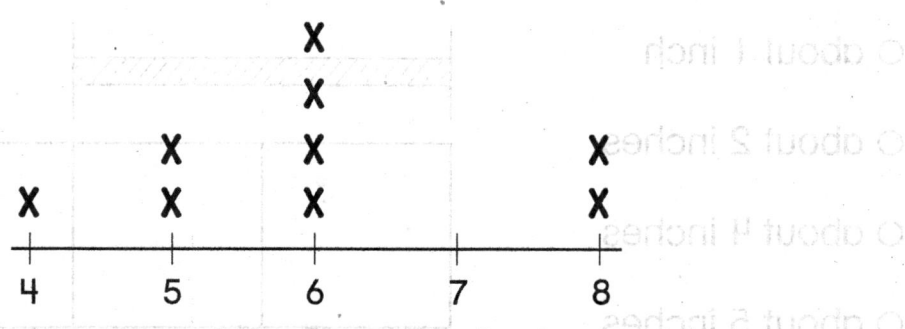

```
              X
              X
        X     X           X
  X     X     X           X
  +-----+-----+-----+-----+
  4     5     6     7     8
```
Lengths of Toy Boats in Inches

1. Suppose one of the toy boats that is 5 inches long breaks and is thrown away. How many 5-inch boats should the line plot now show?

   ○ 0   ○ 1   ○ 2   ○ 3

2. How many toy boats are 4 inches long?

   ○ 1   ○ 2   ○ 4   ○ 6

3. Mia measures the length of a book to the nearest inch. It is about 12 inches long. Which is the same as 12 inches?

   ○ 1 foot

   ○ 2 feet

   ○ 6 feet

   ○ 12 feet

4. Which sentence is **most** likely to be **true**?

   ○ The boy is 40 feet tall.

   ○ The building is 30 feet tall.

   ○ The street is 12 inches wide.

   ○ The driveway is 35 inches long.

   GO ON

Assessment Guide   AG93   Chapter 8 Test
© Houghton Mifflin Harcourt Publishing Company

Name _____

Chapter 8 Test
Page 2

5. Each tile is about 1 inch long. Which is the best choice for the length of the ribbon?

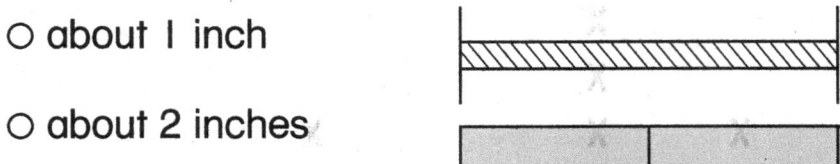

   ○ about 1 inch
   ○ about 2 inches
   ○ about 4 inches
   ○ about 5 inches

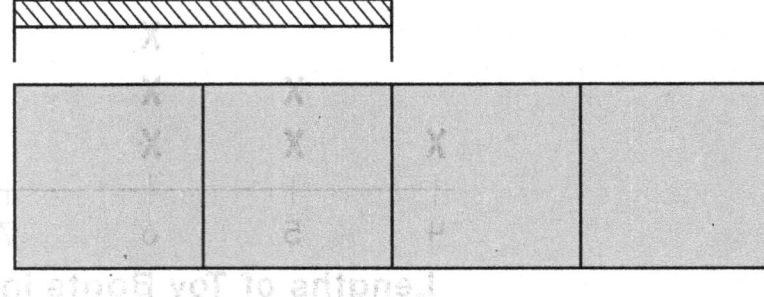

6. Owen made a paper clip chain that was 18 inches long. Then he removed 9 inches of paper clips from the chain. How long is the paper clip chain now?

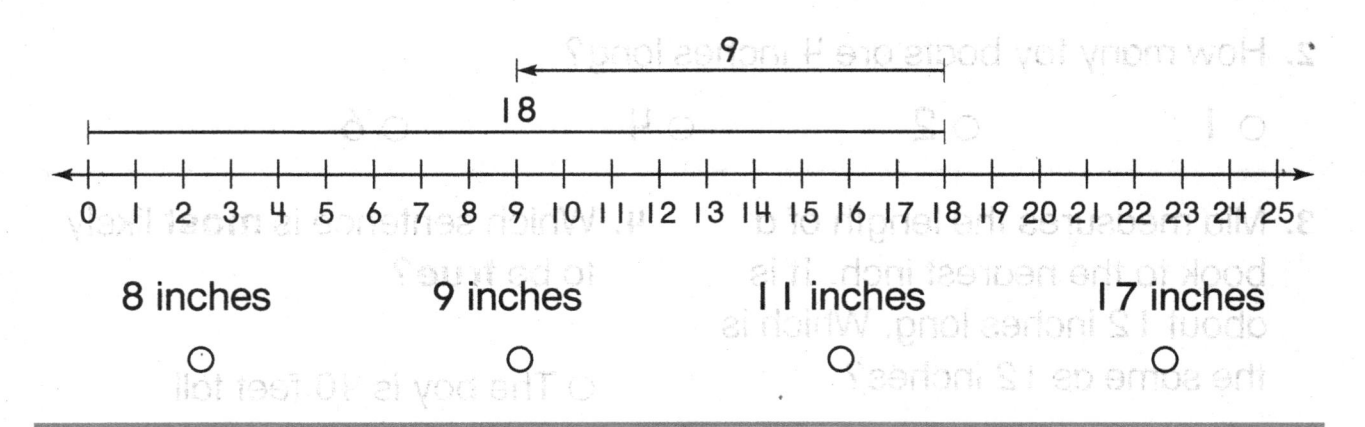

   8 inches   9 inches   11 inches   17 inches
      ○   ○   ○   ○

7. The bead is 1 inch long. What is the best estimate for the length of the string?

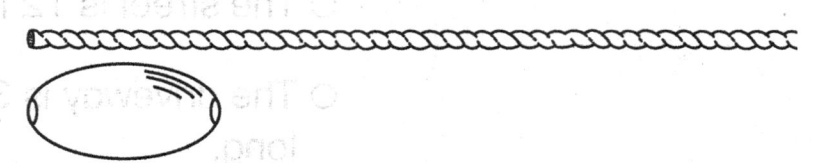

   1 inch   4 inches   7 inches   10 inches
      ○   ○   ○   ○

GO ON

Assessment Guide
© Houghton Mifflin Harcourt Publishing Company

AG94

Chapter 8 Test

Name _____

Chapter 8 Test
Page 3

8. Use an inch ruler. What is the length of the lip balm to the nearest inch?

Lip balm

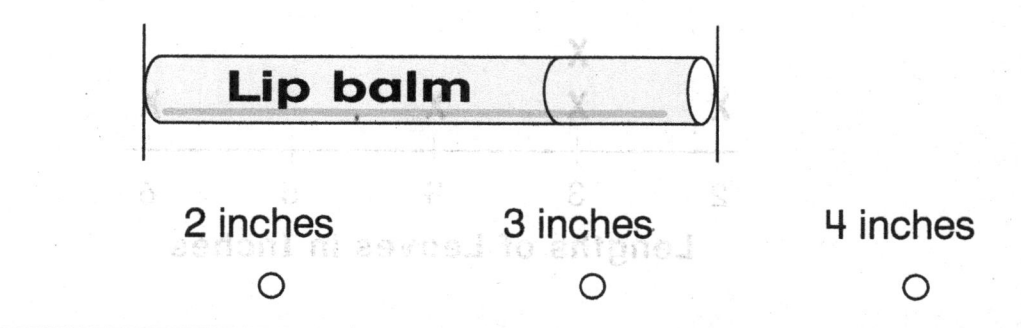

○ 1 inch   ○ 2 inches   ○ 3 inches   ○ 4 inches

9. Use an inch ruler. What is the length of the string to the nearest inch?

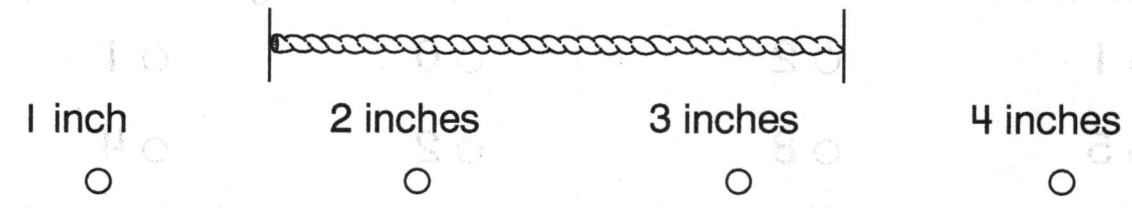

○ 1 inch   ○ 2 inches   ○ 3 inches   ○ 4 inches

10. Which is the **best** estimate for the width of a real classroom door using 1-foot tiles?

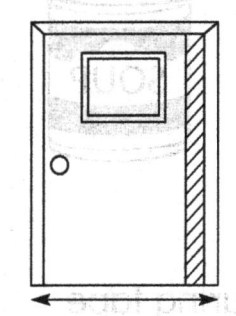

○ 1 tile
○ 3 tiles
○ 8 tiles
○ 20 tiles

11. Use your ruler.

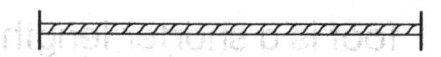

Which is the best choice for the length of the yarn?

○ about 2 inches
○ about 3 inches
○ about 4 inches
○ about 5 inches

GO ON

Assessment Guide
© Houghton Mifflin Harcourt Publishing Company

AG95

Chapter 8 Test

Name _____

Chapter 8 Test
Page 4

Use the line plot for Questions 12–13.

```
              X
    X    X         X              X
  --+----+----+----+----+--
    2    3    4    5    6
       Lengths of Leaves in Inches
```

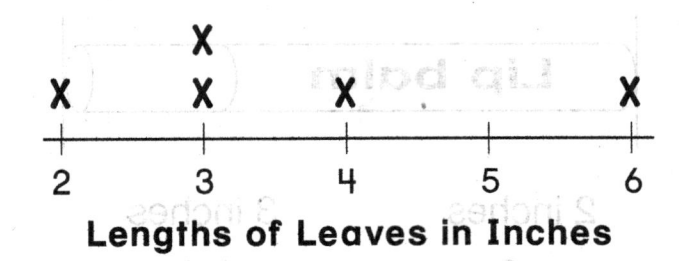

12. How many leaves does the line plot show?
    - ○ 1
    - ○ 2
    - ○ 5
    - ○ 8

13. How many leaves are 5 inches long?
    - ○ 0
    - ○ 1
    - ○ 2
    - ○ 4

14. Leah shows her friend how to use a ruler to measure length. Which sentence is true?

    - ○ 1 foot is a shorter length than 1 inch.
    - ○ 1 inch is a shorter length than 1 foot.
    - ○ 1 inch is the same length as 1 foot.
    - ○ Inches are not used to measure length.

15. Sam wants to measure the distance around a can of soup. Which is the best tool for Sam to use?

    - ○ cup
    - ○ measuring tape
    - ○ large paper clip
    - ○ pencil

GO ON

Assessment Guide
© Houghton Mifflin Harcourt Publishing Company

AG96

Chapter 8 Test

Name _____

Chapter 8 Test
Page 5

16. Each tile is about 1 inch long. Which is the best choice for the length of the straw?

   ○ about 1 inch
   ○ about 3 inches
   ○ about 4 inches
   ○ about 5 inches

17. Lin has a string that is 6 inches long and a string that is 11 inches long. How many inches of string does Lin have altogether?

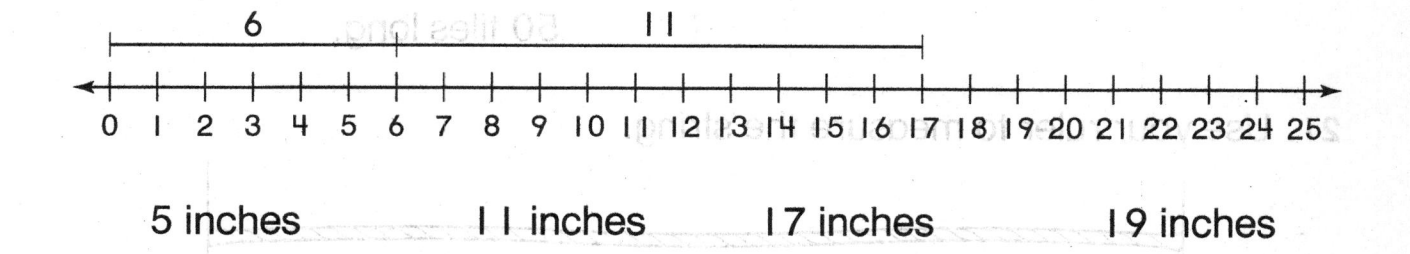

   5 inches   11 inches   17 inches   19 inches
      ○   ○   ○   ○

18. The bead is 1 inch long. What is the best estimate for the length of the string?

   1 inch   3 inches   5 inches   7 inches
     ○   ○   ○   ○

GO ON

Assessment Guide
© Houghton Mifflin Harcourt Publishing Company

Name _____

Chapter 8 Test
Page 6

19. Use an inch ruler. What is the length of the crayon to the nearest inch?

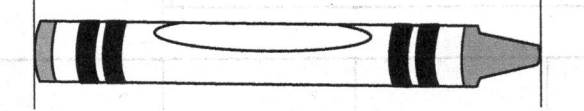

○ 2 inches
○ 3 inches
○ 4 inches
○ 5 inches

20. Suppose 1-inch tiles are used to estimate the length of a football. Which sentence is correct?

○ The football is less than 2 tiles long.
○ The football is about 6 tiles long.
○ The football is about 12 tiles long.
○ The football is more than 50 tiles long.

21. Use your ruler to measure the string.

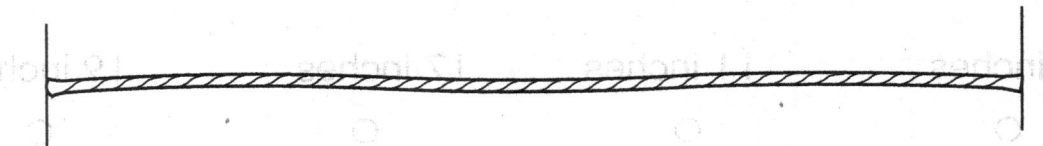

What is the best choice for the length of the string?

○ about 3 inches   ○ about 4 inches
○ about 5 inches   ○ about 6 inches

22. Use your ruler to measure the marker.

What is the best choice for the length of the marker?

○ about 2 inches   ○ about 3 inches
○ about 4 inches   ○ about 5 inches

GO ON

Assessment Guide
© Houghton Mifflin Harcourt Publishing Company

AG98

Chapter 8 Test

Name _____

Chapter 8 Test
Page 7

23. Marta measured some crayons in a box.

Lengths of Crayons	
5 inches	4 inches
2 inches	4 inches
3 inches	

A. USE the data in the list to make a line plot.

Answer:

Name _____

Chapter 8 Test
Page 8

B. Suppose the crayon that is 5 inches long broke. Now it is 3 inches long. **EXPLAIN** how the line plot should be changed.

PUT your answer in the **BLANKS BELOW**.

Answer: _____
_____
_____
_____

C. **COUNT** the crayons that are second longest. How many are there?

PUT your answer in the **BLANK BELOW**.

Answer: _____

Name _____

Chapter 9 Test
Page 1

Choose the correct answer.

1. Which words make the sentence true?

    I centimeter is _____ I meter.

    ○ longer than
    ○ shorter than
    ○ the same as

2. Mandy used unit cubes to measure the length of a straw. Which is the best choice for the length of the straw?

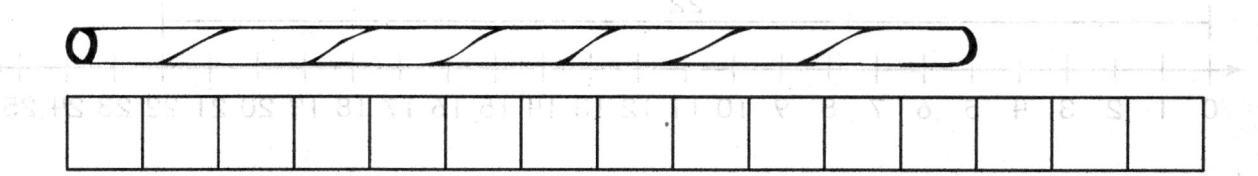

    ○ 12 centimeters   ○ 13 centimeters
    ○ 14 centimeters   ○ 15 centimeters

3. Ken used unit cubes to measure the length of a stick. Which is the best choice for the length of the stick?

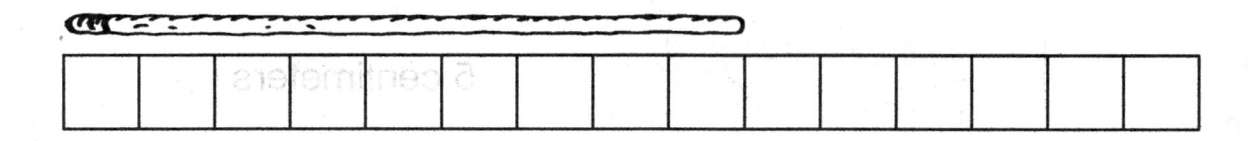

    ○ 6 centimeters   ○ 7 centimeters
    ○ 9 centimeters   ○ 16 centimeters

GO ON

Assessment Guide
© Houghton Mifflin Harcourt Publishing Company

AG101

Chapter 9 Test

Name _____

Chapter 9 Test
Page 2

4. Estimate the length of a real horse. Which is the **best** answer?

  ○ The horse is about 3 meters long.
  ○ The horse is less than 1 meter long.
  ○ The horse is about 6 centimeters long.
  ○ The horse is less than 10 centimeters long.

5. Ms. Diaz had a board that was 22 centimeters long. Then she cut 8 centimeters off the board. How long is the board now?

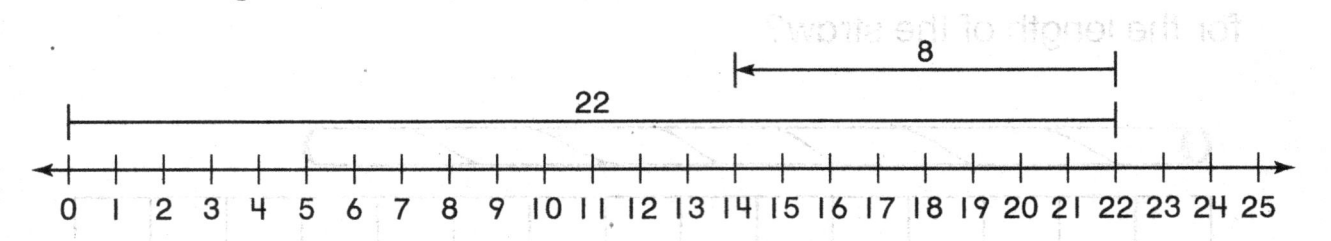

  ○ 8 centimeters   ○ 14 centimeters
  ○ 22 centimeters   ○ 30 centimeters

6. Which number sentence can be used to find how much longer the ribbon is than the paper clip?

  9 centimeters

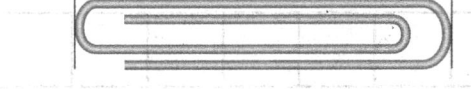

  5 centimeters

  ○ 9 + 5 = 14   ○ 9 − 5 = 4
  ○ 9 + 4 = 13   ○ 5 − 4 = 1

  GO ON

Assessment Guide
© Houghton Mifflin Harcourt Publishing Company

AG102

Chapter 9 Test

Name _____

Chapter 9 Test
Page 3

7. The string is about 2 centimeters long.

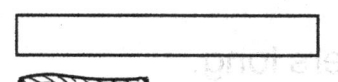

   Which is the **best** estimate for the length of the strip of paper?

   ○ 1 centimeter   ○ 2 centimeters
   ○ 4 centimeters   ○ 6 centimeters

8. The crayon is about 8 centimeters long.

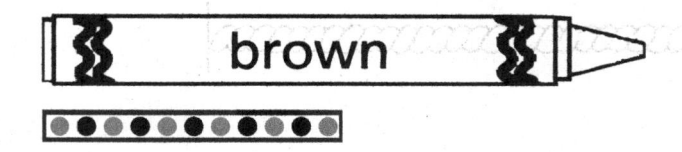

   Which is the **best** estimate for the length of the ribbon?

   ○ 2 centimeters   ○ 4 centimeters
   ○ 6 centimeters   ○ 8 centimeters

9. Use a centimeter ruler. What is the length of the yarn to the nearest centimeter?

   ○ 2 centimeters   ○ 3 centimeters
   ○ 4 centimeters   ○ 5 centimeters

   GO ON

Assessment Guide
© Houghton Mifflin Harcourt Publishing Company

Name _____

Chapter 9 Test
Page 4

10. Which sentence makes the **most** sense?

    ○ A car is 6 centimeters long.

    ○ A sidewalk is 2 meters wide.

    ○ A swimming pool is 50 centimeters long.

    ○ A computer keyboard is 42 meters wide.

11. Use a centimeter ruler. How much longer is the pencil than the string?

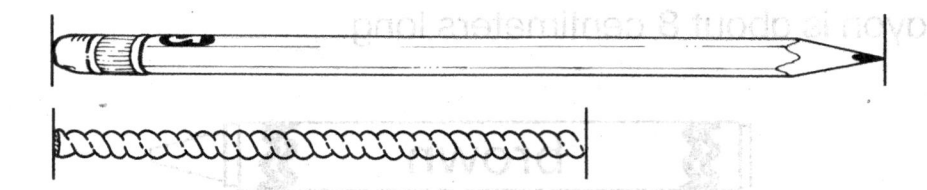

    ○ 4 centimeters longer   ○ 7 centimeters longer

    ○ 11 centimeters longer   ○ 18 centimeters longer

12. Susan uses unit cubes to measure the length of the yarn. What is the **best** estimate for the length of the yarn?

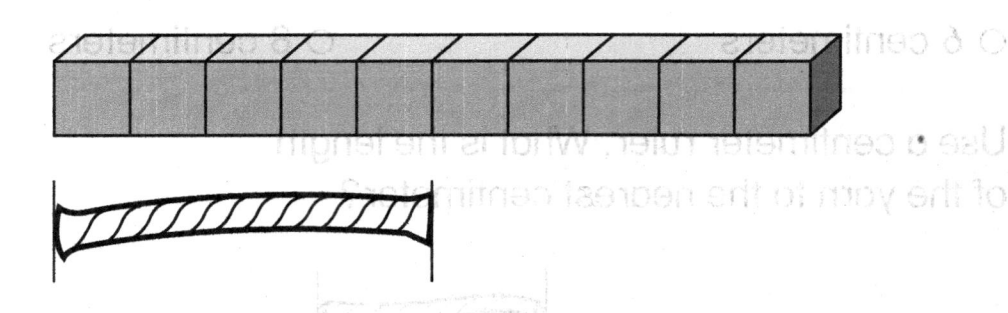

    ○ 2 centimeters   ○ 3 centimeters

    ○ 5 centimeters   ○ 9 centimeters

GO ON

Assessment Guide
© Houghton Mifflin Harcourt Publishing Company

AG104

Chapter 9 Test

Name _____

Chapter 9 Test
Page 5

13. Which is the **best** estimate for the length of a window?

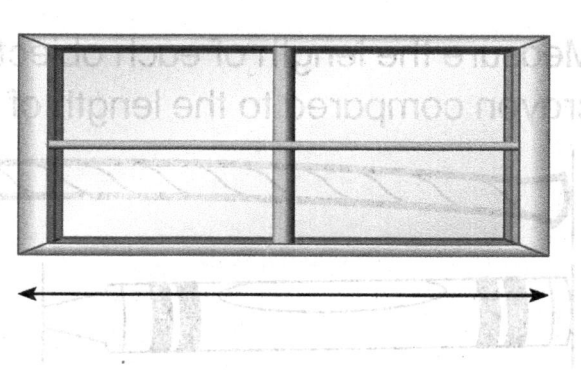

   ○ 2 meters   ○ 5 meters   ○ 7 meters   ○ 10 meters

14. Ann has a rope that is 12 centimeters long. Hiro has a rope that is 5 centimeters long. How many centimeters of rope do they have altogether?

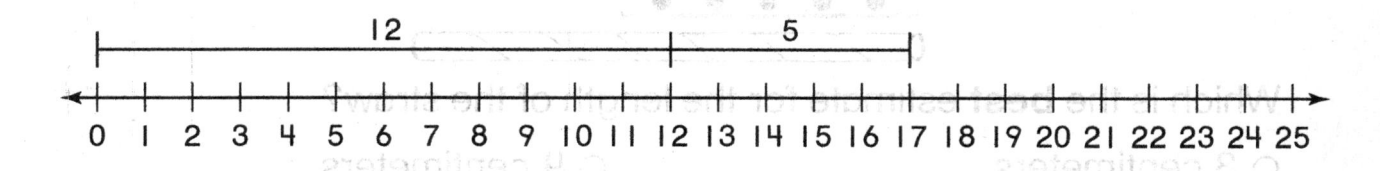

   ○ 7 centimeters   ○ 12 centimeters
   ○ 16 centimeters   ○ 17 centimeters

15. Roy drew a mark that was 15 centimeters long. Then he erased 6 centimeters from the mark. How long is the mark now?

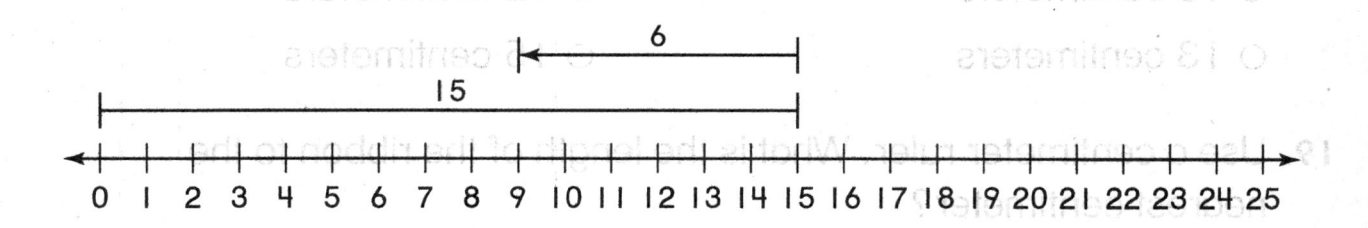

   ○ 6 centimeters   ○ 9 centimeters
   ○ 15 centimeters   ○ 21 centimeters

GO ON

Assessment Guide
© Houghton Mifflin Harcourt Publishing Company

AG105

Chapter 9 Test

Name _____

Chapter 9 Test
Page 6

16. Measure the length of each object. Which describes the length of the crayon compared to the length of the yarn?

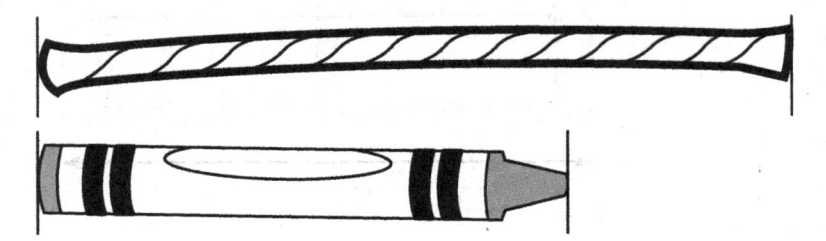

○ 3 centimeters shorter ○ 3 centimeters longer

○ 7 centimeters shorter ○ 7 centimeters longer

17. The ribbon is about 4 centimeters long.

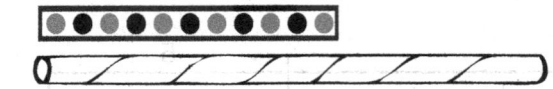

Which is the **best** estimate for the length of the straw?

○ 3 centimeters ○ 4 centimeters

○ 7 centimeters ○ 9 centimeters

18. Measure the length of the rope to the nearest centimeter. What is the **best** estimate for the length of the rope?

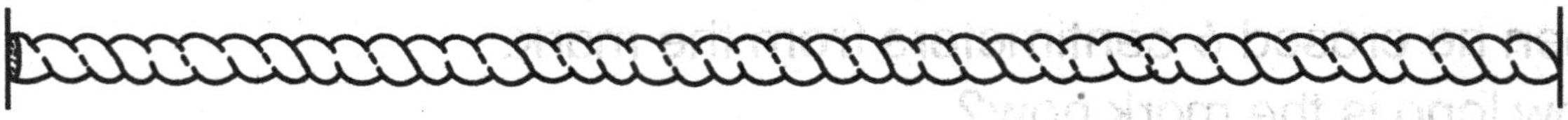

○ 10 centimeters ○ 12 centimeters

○ 13 centimeters ○ 15 centimeters

19. Use a centimeter ruler. What is the length of the ribbon to the nearest centimeter?

○ 5 centimeters ○ 6 centimeters

○ 7 centimeters ○ 8 centimeters

GO ON

Assessment Guide
© Houghton Mifflin Harcourt Publishing Company

AG106

Chapter 9 Test

Name _____

Chapter 9 Test
Page 7

20. Alexis has these three pieces of string left over from wrapping up bundles of wood. Each cube is 1 cm long.

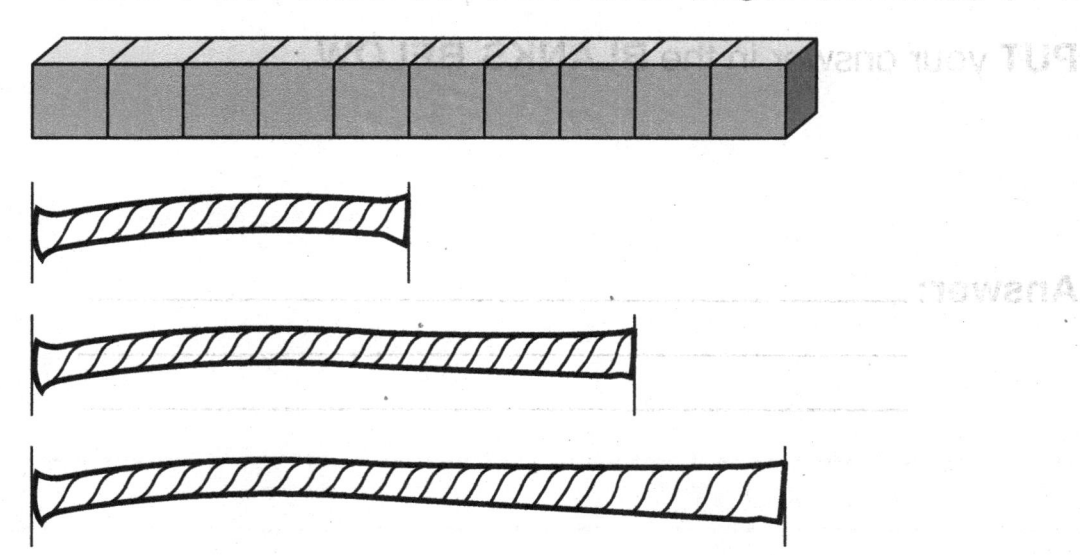

A. What are the lengths of each piece of string?
PUT your answer in the **BLANK BELOW**.

Answer: _____

Name _____

Chapter 9 Test
Page 8

B. CHOOSE the two pieces of string that have a total length of 13 centimeters. **EXPLAIN** how you found your answer.

PUT your answer in the **BLANKS BELOW**.

Answer: _____
_____
_____

C. ADD to find the total length of the longest piece of string plus the shortest piece of string.

PUT your answer in the **BLANK BELOW**.

Answer: _____

Assessment Guide
© Houghton Mifflin Harcourt Publishing Company

AG108

Chapter 9 Test

Name _____

Chapter 10 Test
Page 1

Choose the correct answer.

Use the bar graph for Questions 1–4.

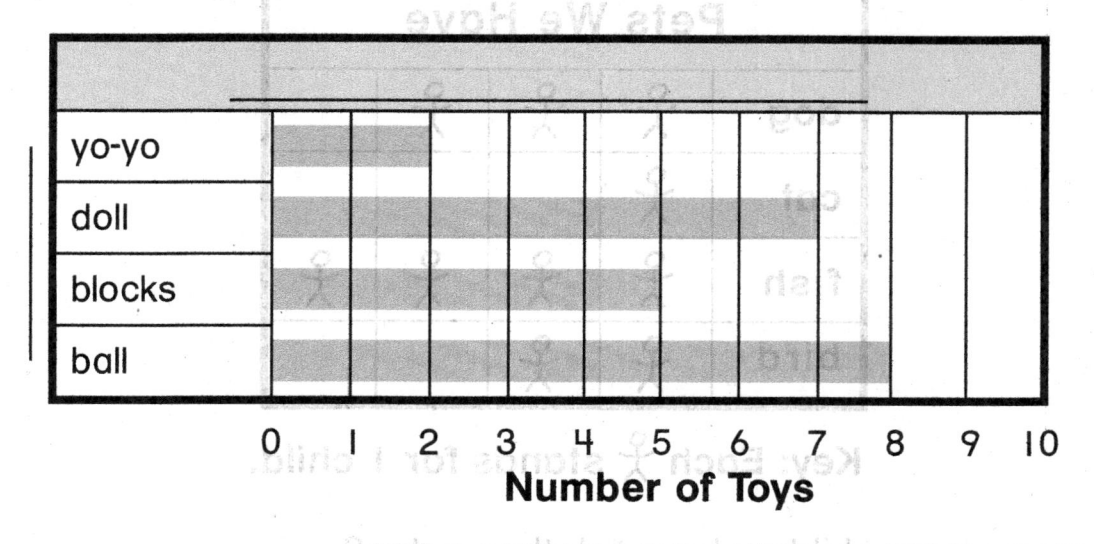

	0	1	2	3	4	5	6	7	8	9	10
yo-yo											
doll											
blocks											
ball											

Number of Toys

1. Which is the **best** label for the rows in the bar graph?

   ○ Toy ○ Children

   ○ Yo-Yos ○ Game

2. Which is the **best** title for the bar graph?

   ○ Books We Like ○ Children's Dolls

   ○ Favorite Yo-Yo ○ Toys in the Store

3. How many more dolls than blocks are in the store?

   ○ 1 ○ 2

   ○ 5 ○ 7

4. How many balls are in the store?

   ○ 5 ○ 7

   ○ 8 ○ 10

GO ON ➡

Assessment Guide
© Houghton Mifflin Harcourt Publishing Company

AG109

Chapter 10 Test

Name _____

Chapter 10 Test
Page 2

Use the picture graph for Questions 5–8.

Pets We Have				
dog	☺	☺	☺	
cat	☺			
fish	☺	☺	☺	☺
bird	☺	☺		

Key: Each ☺ stands for 1 child.

5. How many more children have fish than a dog?
   ○ 0   ○ 1   ○ 2   ○ 3

6. What are two animals that a total of 6 children have?
   ○ dog and cat   ○ bird and cat
   ○ fish and dog   ○ fish and bird

7. **2 more** children have cats. Now how many ☺ should be in the cat row of the picture graph?
   ○ 3
   ○ 4
   ○ 5
   ○ 6

8. 3 children have a hamster. Which shows a hamster row for the picture graph?
   ○ ☺ ☺ ☺ ☺ ☺
   ○ ☺ ☺ ☺ ☺
   ○ ☺ ☺ ☺
   ○ ☺ ☺

GO ON

Name _____

Chapter 10 Test
Page 3

Use the tally chart for Questions 9–12.

Favorite Month	
Month	Tally
May	⊞
June	⊞ I
July	IIII
August	III

9. Which statement is **true**?

   ○ More than 10 children like July or August.

   ○ June is the least favorite month.

   ○ More children like May than August.

   ○ Most children like July.

10. Which month did the **most** children choose?

    ○ May
    ○ June
    ○ July
    ○ August

11. Which month did the **fewest** children choose?

    ○ May   ○ July
    ○ June   ○ August

12. **7 more** children were asked their favorite month. 6 children like May the **most** and 1 child likes August the **most**. Now which month is chosen as favorite by the **most** children?

    ○ May   ○ July
    ○ June   ○ August

GO ON

Assessment Guide
© Houghton Mifflin Harcourt Publishing Company

AG111

Chapter 10 Test

Name _____

Chapter 10 Test
Page 4

13. Joe read for 5 hours in Week 1, 3 hours in Week 2, and 2 hours in Week 3. Which bar graph shows this data?

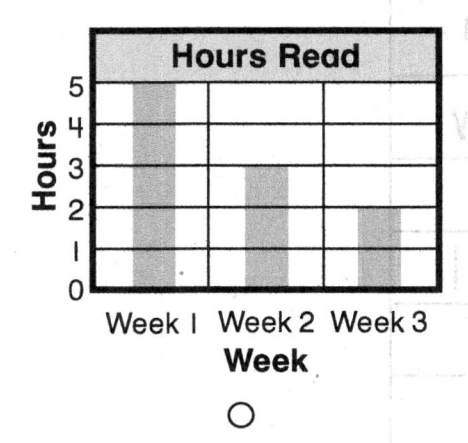

○   ○

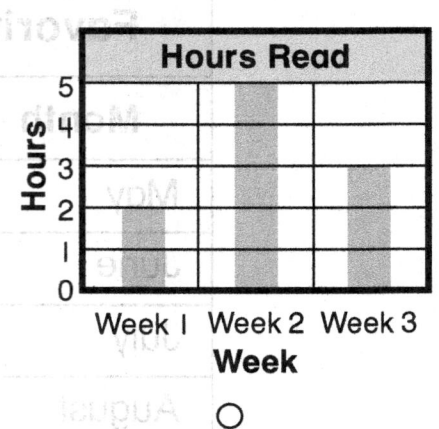

Use the bar graph for Questions 14–16.

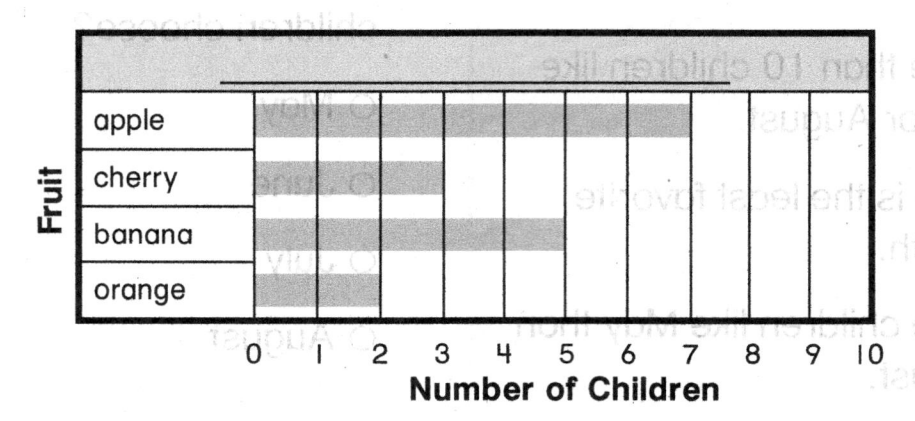

14. Which could be the title of the bar graph?

    ○ Our Pets   ○ Apples   ○ Favorite Fruit   ○ Books Read

15. How many **more** children chose banana or cherry than apple?

    ○ 1   ○ 2   ○ 3   ○ 4

16. Which of the following sentences is **true**?

    ○ 6 children chose banana.
    ○ 7 children chose cherry or orange.
    ○ Apple was the most popular.
    ○ 3 more children chose banana than cherry.

GO ON ➡

Assessment Guide
© Houghton Mifflin Harcourt Publishing Company

AG112

Chapter 10 Test

Name _____

Chapter 10 Test
Page 5

Use the picture graph for Questions 17–20.

Favorite Sports					
hockey	☺	☺			
football	☺	☺	☺		
soccer	☺	☺	☺	☺	☺
basketball	☺	☺	☺		

Key: Each ☺ stands for 1 child.

17. How many children chose soccer?
    ○ 2   ○ 3
    ○ 4   ○ 5

18. How many children in all chose football or hockey?
    ○ 2   ○ 3
    ○ 5   ○ 8

19. 1 more child chose basketball. Now how many ☺ should be in the basketball row of the picture graph?
    ○ 2
    ○ 4
    ○ 5
    ○ 6

20. 4 children like baseball the best. Which shows a baseball row for the picture graph?
    ○ ☺ ☺ ☺ ☺
    ○ ☺ ☺ ☺
    ○ ☺ ☺
    ○ ☺

GO ON

Assessment Guide　　AG113　　Chapter 10 Test
© Houghton Mifflin Harcourt Publishing Company

Name _____

Chapter 10 Test
Page 6

Use the tally chart for Questions 21–24.

Favorite Animal	
Animal	Tally
lion	llll
bear	lll
monkey	ll
giraffe	llll ll

21. Which statement is **true**?

○ 8 children chose giraffe.

○ 17 children voted in all.

○ 5 children voted for bear.

○ More children chose monkey than lion.

22. Which animal did the **fewest** children choose?

○ lion

○ bear

○ monkey

○ giraffe

23. Which animal did the **most** children choose?

○ lion   ○ monkey

○ bear   ○ giraffe

24. How many children chose lion?

○ 3   ○ 4

○ 5   ○ 6

GO ON

Assessment Guide
© Houghton Mifflin Harcourt Publishing Company

AG114

Chapter 10 Test

Name _____

Chapter 10 Test
Page 7

25. Nathan asked his friends to choose their favorite sport. He wrote the results in the picture graph below.

Favorite Sports					
hockey	☺	☺			
football	☺	☺	☺		
soccer	☺	☺	☺	☺	☺
basketball	☺	☺	☺		

Key: Each ☺ stands for 1 child.

A. Nathan asks 3 more friends which sport they like best. 2 friends choose football and 1 friend chooses hockey. Which sport is the most chosen now?

EXPLAIN your answer in the **BLANKS BELOW**.

Answer: _____
_____
_____

Name _____

**Chapter 10 Test
Page 8**

B. How many children are there in all now?

PUT your answer in the **BLANK BELOW**.

Answer: _____ children

C. Which sport was the **least** popular? **EXPLAIN** your answer in the **BLANKS BELOW**.

Answer: _____

_____

Name _____

Chapter 11 Test
Page 1

Choose the correct answer.

1. What is the shape of this sign?

   ○ triangle
   ○ quadrilateral
   ○ pentagon
   ○ hexagon

2. Alex built this rectangular prism. How many unit cubes did Alex use?

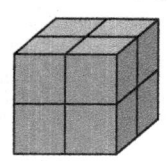

   ○ 7
   ○ 8
   ○ 16
   ○ 18

3. Which shape has **fewer** than 4 angles?

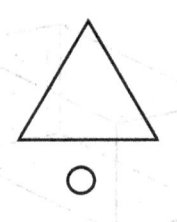

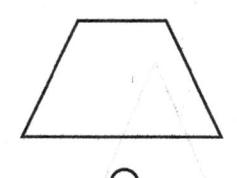

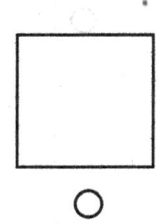

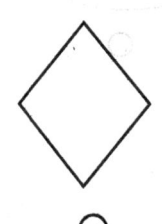

   ○   ○
   ○   ○

4. Lisa used square tiles to cover this rectangle. How many square tiles did she use to cover the rectangle?

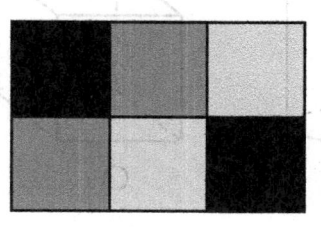

   ○ 4
   ○ 6
   ○ 8
   ○ 9

GO ON

Assessment Guide AG117 Chapter 11 Test

Name _____

Chapter 11 Test
Page 2

5. Paul makes a hexagon and a triangle with straws. He uses one straw for each side of a shape. How many straws does Paul need?

   ○ 8
   ○ 9
   ○ 10
   ○ 11

6. Which shape shows fourths?

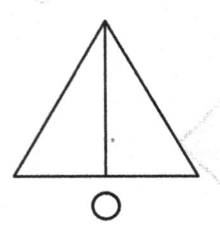

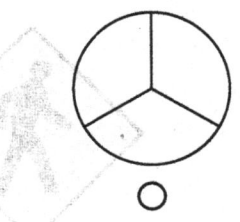

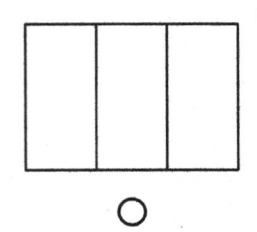

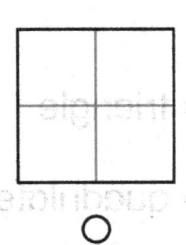

7. Which of these shapes is a cube?

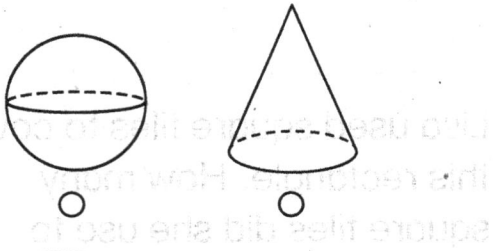

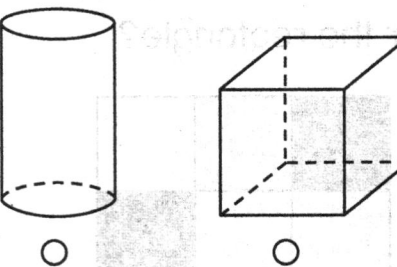

8. Which of these shapes is a cylinder?

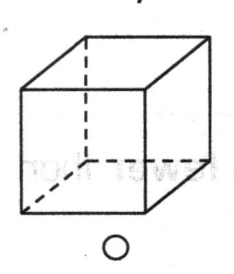

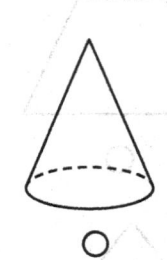

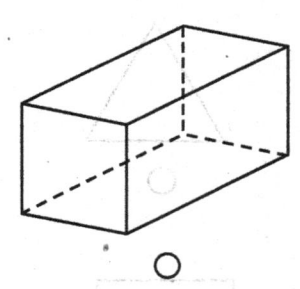

Assessment Guide
© Houghton Mifflin Harcourt Publishing Company

AG118

GO ON ➡

Chapter 11 Test

Name _____

Chapter 11 Test
Page 3

9. Which shape has parts that are halves?

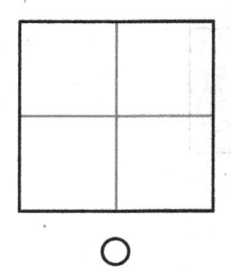

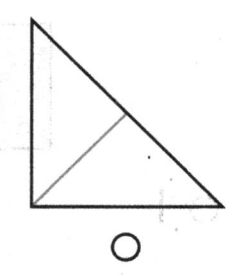

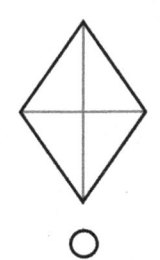

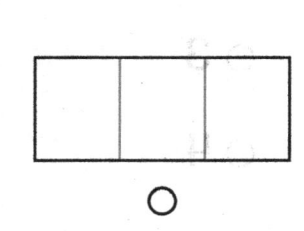

10. Which shape shows thirds?

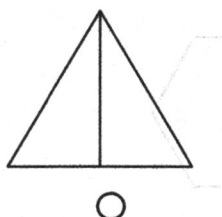

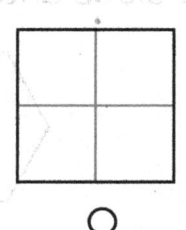

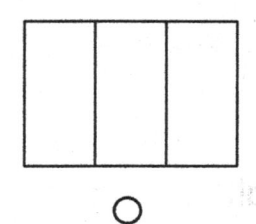

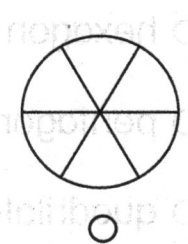

11. Which sentence does **not** describe a rectangular prism?

○ A rectangular prism has 16 vertices.

○ A rectangular prism has 12 edges.

○ A rectangular prism has 6 faces.

○ At least two faces of a rectangular prism are rectangles.

12. Which shape shows 4 equal parts?

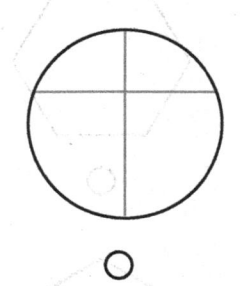

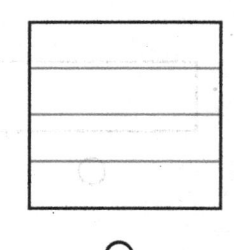

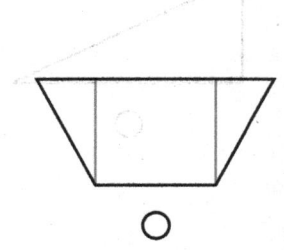

GO ON

Assessment Guide
© Houghton Mifflin Harcourt Publishing Company

AG119

Chapter 11 Test

Name _____

**Chapter 11 Test
Page 4**

13. Which names a shape with 6 sides and 6 vertices?

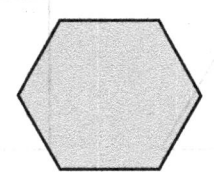

    ○ hexagon

    ○ pentagon

    ○ quadrilateral

    ○ triangle

14. How many sides does this quadrilateral have?

    ○ 1

    ○ 2

    ○ 3

    ○ 4

15. Which of these shapes has more than 5 angles?

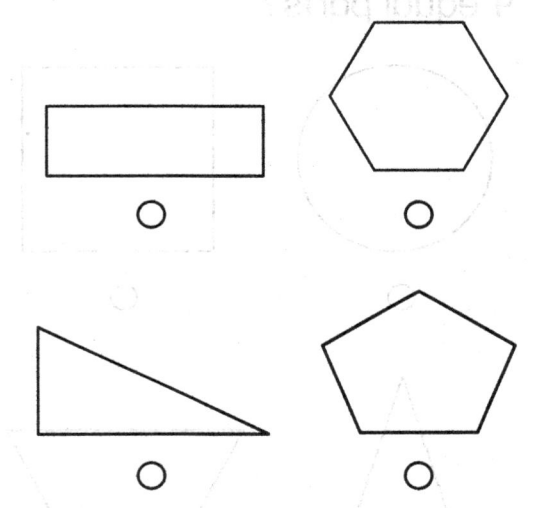

16. Theo used square tiles to cover this rectangle. How many square tiles did he use to cover the rectangle?

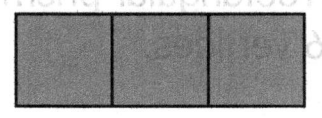

    ○ 1

    ○ 2

    ○ 3

    ○ 4

GO ON

Assessment Guide
© Houghton Mifflin Harcourt Publishing Company

AG120

Chapter 11 Test

Name _____

Chapter 11 Test
Page 5

17. How many angles does this shape have?

○ 3
○ 4
○ 5
○ 6

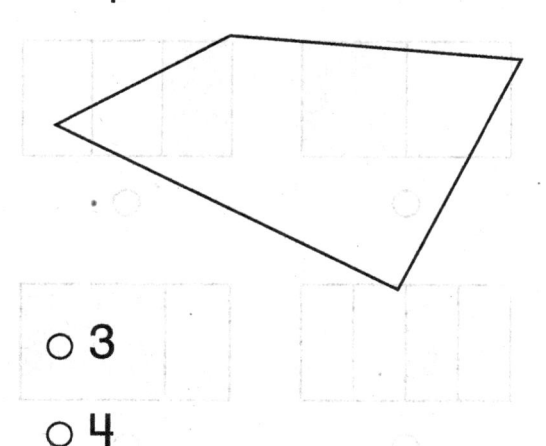

18. Paul cuts a sheet of paper into thirds like this.

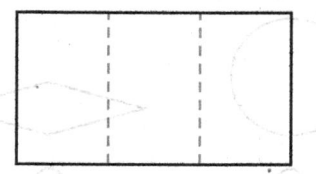

Which shows another way to cut the paper into thirds?

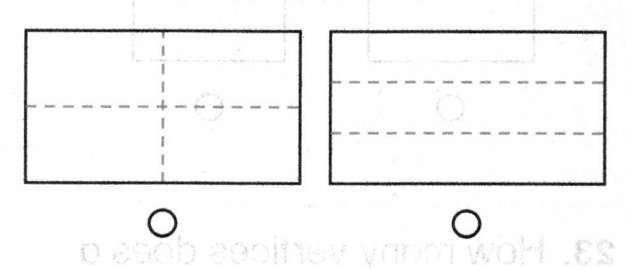

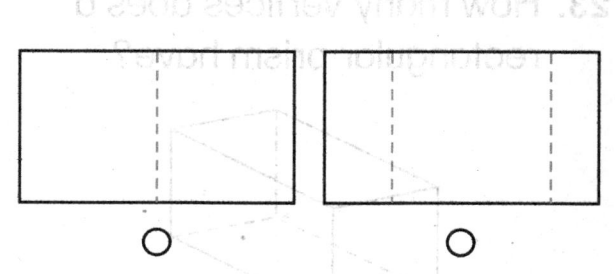

○   ○

○   ○

19. Which object is shaped like a sphere?

○   ○

○   ○

20. Which object is shaped like a rectangular prism?

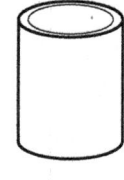

○   ○

○   ○

GO ON

Assessment Guide
© Houghton Mifflin Harcourt Publishing Company

AG121

Chapter 11 Test

Name _____

Chapter 11 Test
Page 6

21. Which shape has parts that are thirds?

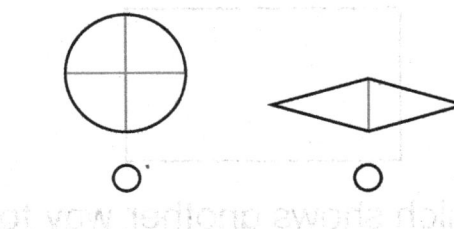

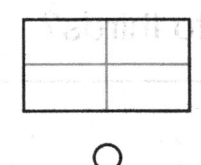

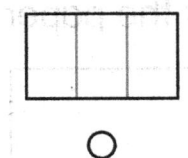

○   ○

○   ○

22. Which shows a half of the shape shaded?

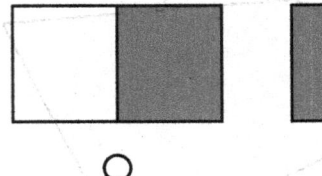

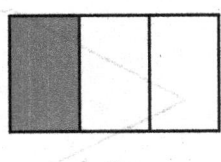

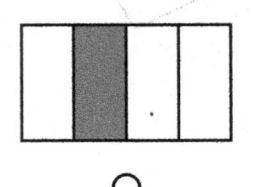

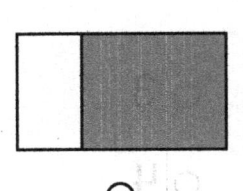

○   ○

○   ○

23. How many vertices does a rectangular prism have?

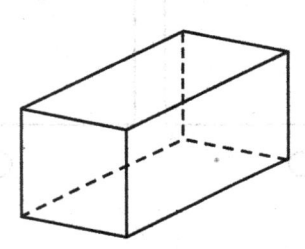

○ 4
○ 5
○ 6
○ 8

24. Which shape shows halves?

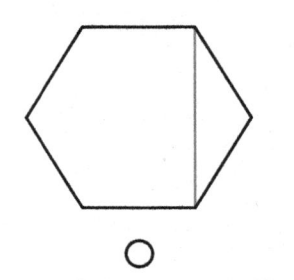

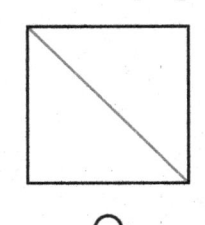

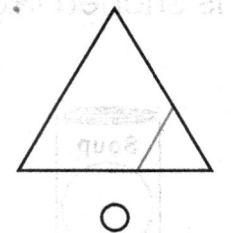

○   ○

○   ○

GO ON ➡

Assessment Guide
© Houghton Mifflin Harcourt Publishing Company

AG122

Chapter 11 Test

Name _____

Chapter 11 Test
Page 7

25. Mason drew 2 two-dimensional shapes that had 8 angles in all.

A. DRAW the shapes Mason could have drawn on the **GRID BELOW**.

Name _____

Chapter 11 Test
Page 8

B. **NAME** the shapes.

   PUT your answer in the **BLANK BELOW**.

   Answer: _____

C. **NAME** the shapes Mason could draw if he drew 2 two-dimensional shapes with 7 angles in all.

   PUT your answer in the **BLANK BELOW**.

   Answer: _____
   _____

Child's Name _____ Date _____

Beginning-of-Year/Middle-of-Year/End-of-Year Test

Item	Lesson	PA Common Core Standard	Common Error	Intervene with	Personal Math Trainer
1	3.5	CC.2.2.2.A.2	May not understand the term *related fact*	R—3.5	2.OA.2
2	3.9	CC.2.2.2.A.2	May use an incorrect number sentence to solve	R—3.9	2.OA.1
3	3.4	CC.2.2.2.A.2	May not add three addends correctly	R—3.4	2.OA.2
4	6.8	CC.2.1.2.B.3	May not reduce the number in the hundreds column after regrouping	R—6.8	2.NBT.7
5	6.3	CC.2.1.2.B.3	May not regroup correctly when adding the ones or tens	R—6.3	2.NBT.7
6	6.10	CC.2.1.2.B.3	May not regroup correctly when there is a 0 in the tens	R—6.10	2.NBT.7
7	8.4	CC.2.4.2.A.1	May not line up the edge of the ruler when measuring	R—8.4	2.MD.1
8	8.9	CC.2.4.2.A.4	May have difficulty reading a line plot	R—8.9	2.MD.9
9	8.7	CC.2.4.2.A.1	May have difficulty estimating length in feet	R—8.7	2.MD.3
10	8.8	CC.2.4.2.A.1	May not understand the purposes of different measuring tools	R—8.8	2.MD.1
11	1.1	CC.2.2.2.A.3	May not understand the meaning of *even* and *odd*	R—1.1	2.OA.3
12	1.3	CC.2.1.2.B.2	May not know the value of a digit in the ones or tens place	R—1.3	2.NBT.3
13	1.5	CC.2.1.2.B.2	May not understand that a number can be written in different ways	R—1.5	2.NBT.3
14	1.9	CC.2.1.2.B.2	May not understand how to count by 10s.	R—1.9	2.NBT.2
15	11.1	CC.2.3.2.A.1	May not be able to identify three-dimensional shapes	R—11.1	2.G.1
16	11.4	CC.2.3.2.A.1	May not understand properties of shapes.	R—11.4	2.G.1
17	11.5	CC.2.3.2.A.1	May not understand how to sort shapes according to the number of sides and vertices	R—11.5	2.G.1

Key: R—Reteach

Assessment Guide
© Houghton Mifflin Harcourt Publishing Company

Individual Record Form

Child's Name _____   Date _____

Beginning-of-Year/Middle-of-Year/End-of-Year Test

Item	Lesson	PA Common Core Standard	Common Error	Intervene with	Personal Math Trainer
18	11.9	CC.2.3.2.A.2	May not be able to identify a half, a third, or a fourth of a shape	R—11.9	2.G.3
19	4.10	CC.2.2.2.A.1	May find a difference instead of a sum	R—4.10	2.OA.1
20	4.11	CC.2.1.2.B.2	May forget to add the third addend	R—4.11	2.NBT.6
21	4.9	CC.2.1.2.B.2	May add incorrectly	R—4.9	2.NBT.5
22	4.7	CC.2.1.2.B.2	May add incorrectly	R—4.7	2.NBT.5
23	10.2	CC.2.4.2.A.4	May misread the picture graph	R—10.2	2.MD.10
24	10.3	CC.2.4.2.A.4	May not understand how to complete a row of a picture graph	R—10.3	2.MD.10
25	10.1	CC.2.4.2.A.4	May not know that there are 5 tallies in each bundle	R—10.1	2.MD.10
26	10.4	CC.2.4.2.A.4	May misread the bar graph	R—10.4	2.MD.10
27	2.5	CC.2.1.2.B.2	May not correctly identify the place value of the digits	R—2.5	2.NBT.1
28	2.7	CC.2.1.2.B.2	May not recognize a number in expanded form	R—2.7	2.NBT.3
29	2.10	CC.2.1.2.B.2	May not continue the pattern correctly	R—2.10	2.NBT.8
30	2.12	CC.2.1.2.B.2	May not know how to use the <, >, and = symbols	R—2.12	2.NBT.4
31	7.2	CC.2.1.2.B.3	May not be able to determine the value of a collection of coins	R—7.2	2.MD.8
32	7.5	CC.2.1.2.B.3	May not be able to identify coins that have a value of one dollar	R—7.5	2.MD.8
33	7.11	CC.2.1.2.B.3	May not understand A.M. and P.M.	R—7.11	2.MD.7
34	5.1	CC.2.1.2.B.3	May break apart the ones incorrectly	R—5.1	2.NBT.5

Key: R—Reteach

Assessment Guide
© Houghton Mifflin Harcourt Publishing Company

Individual Record Form

Child's Name _____ Date _____

Beginning-of-Year/Middle-of-Year/End-of-Year Test

Item	Lesson	PA Common Core Standard	Common Error	Intervene with	Personal Math Trainer
35	5.10	CC.2.2.2.A.2	May not understand how to write a number sentence to represent the problem	R—5.10	2.OA.1
36	5.11	CC.2.2.2.A.2	May forget to complete all the steps to solve the problem	R—5.11	2.OA.1
37	5.5	CC.2.1.2.B.3	May not subtract correctly	R—5.5	2.NBT.5
38	9.5	CC.2.4.2.A.1	May not understand the relationship between a centimeter and a meter	R—9.5	2.MD.2
39	9.3	CC.2.4.2.A.1	May not line up the end of the object with the 0 mark on the centimeter ruler	R—9.3	2.MD.1
40	9.7	CC.2.4.2.A.1	May perform the wrong operation when solving a problem about comparing lengths	R—9.7	2.MD.4
41	10.5	CC.2.4.2.A.4	May not draw the bar graph correctly	R—10.5	2.MD.10

Key: R—Reteach

Child's Name _____ Date _____

Chapter 1 Test

Item	Lesson	PA Common Core Standard	Content Focus	Intervene With	Personal Math Trainer
19–21	1.1	CC.2.2.2.A.3	Classify numbers as even or odd.	R—1.1	2.OA.3
1, 2	1.2	CC.2.2.2.A.3	Represent an even number.	R—1.2	2.OA.3
22–24	1.3	CC.2.1.2.B.2	Identify the value of a digit in a 2-digit number.	R—1.3	2.NBT.3
11, 12	1.9	CC.2.1.2.B.2	Count by 10s.	R—1.9	2.NBT.2
3, 4	1.8	CC.2.1.2.B.2	Count by 2s and 5s.	R—1.8	2.NBT.2
5–7, 25	1.4	CC.2.1.2.B.2	Represent a 2-digit number with a drawing and describe in different forms.	R—1.4	2.NBT.3
13–15	1.7	CC.2.1.2.B.2	Use understanding of place value to solve problems.	R—1.7	2.NBT.3
16–18	1.6	CC.2.1.2.B.2	Use understanding of place value to solve problems.	R—1.6	2.NBT.3
8–10	1.5	CC.2.1.2.B.1	Write 2-digit numbers in word form, expanded form, and standard form.	R—1.5	2.NBT.1

Key: **R**—Reteach

Assessment Guide
© Houghton Mifflin Harcourt Publishing Company

Individual Record Form

Child's Name _____ Date _____

Chapter 2 Test

Item	Lesson	PA Common Core Standard	Content Focus	Intervene With	Personal Math Trainer
13, 14	2.1	CC.2.1.2.B.1	Identify 10 tens as equivalent to 100.	R—2.1	2.NBT.1.a, 2.NBT.1.b
1, 2	2.2	CC.2.1.2.B.1	Use groups of tens to solve problems with 3-digit numbers.	R—2.2	2.NBT.1
3, 4	2.9	CC.2.1.2.B.3	Identify 10 more, 100 less.	R—2.9	2.NBT.8
17, 18	2.10	CC.2.1.2.B.3	Use place value to identify and extend counting patterns.	R—2.10	2.NBT.8
21, 22	2.12	CC.2.1.2.B.1	Compare 3-digit numbers using >, =, and <.	R—2.12	2.NBT.4
5	2.4	CC.2.1.2.B.2	Write 3-digit numbers in word form and expanded form.	R—2.4	2.NBT.3
6, 11, 12	2.7	CC.2.1.2.B.2	Write 3-digit numbers in expanded and standard form.	R—2.7	2.NBT.3
23, 24	2.3	CC.2.1.2.B.1	Use place value to identify the values of digits.	R—2.3	2.NBT.1
15, 16	2.8	CC.2.1.2.B.2	Use a model to represent 3-digit numbers.	R—2.8	2.NBT.3
7, 8, 25	2.11	CC.2.1.2.B.1	Use a model to solve problems using number comparisons.	R—2.11	2.NBT.4
9, 10	2.5	CC.2.1.2.B.1	Use place to identify the values of digits.	R—2.5	2.NBT.1
19, 20	2.6	CC.2.1.2.B.2	Write a 3-digit number in word form.	R—2.6	2.NBT.3

Key: R—Reteach

Assessment Guide
© Houghton Mifflin Harcourt Publishing Company

Individual Record Form

Child's Name _____ Date _____

Chapter 3 Test

Item	Lesson	PA Common Core Standard	Content Focus	Intervene With	Personal Math Trainer
7, 8, 19	3.4	CC.2.2.2.A.2	Add 3 addends.	R—3.4	2.OA.2
4, 16	3.1	CC.2.2.2.A.2	Identify doubles facts.	R—3.1	2.OA.2
3, 15	3.7	CC.2.2.2.A.2	Subtract using a ten.	R—3.7	2.OA.2
6, 18	3.3	CC.2.2.2.A.2	Add and model addition using a ten.	R—3.3	2.OA.2
12, 24	3.11	CC.2.2.2.A.3	Find number of objects in an array using repeated addition.	R—3.11	2.OA.4
5, 17	3.10	CC.2.2.2.A.3	Represent number of objects in equal groups.	R—3.10	2.OA.4
10, 21	3.5	CC.2.2.2.A.2	Represent subtraction problem using a drawing and a number sentence.	R—3.5	2.OA.2
1, 13, 25	3.2	CC.2.2.2.A.2	Represent addition problem using a number sentence.	R—3.2	2.OA.2
2, 14	3.6	CC.2.2.2.A.2	Identify correct subtraction facts.	R—3.6	2.OA.2
9, 20	3.9	CC.2.2.2.A.2	Represent subtraction problem using a number sentence.	R—3.9	2.OA.1
11, 22, 23	3.8	CC.2.1.2.B.3	Use bar models to represent a variety of addition and subtraction situations.	R—3.8	2.NBT.5

Key: R—Reteach

Child's Name _____ Date _____

Chapter 4 Test

Item	Lesson	PA Common Core Standard	Content Focus	Intervene With	Personal Math Trainer
23, 24	4.10	CC.2.2.2.A.2	Use a model and number sentence to solve an addition problem.	R—4.10	2.OA.1
3, 4, 20, 25	4.11	CC.2.1.2.B.3	Add 3 numbers.	R—4.11	2.NBT.6
9, 10	4.6	CC.2.1.2.B.3	Add 2-digit numbers.	R—4.6	2.NBT.5
21, 22	4.5	CC.2.1.2.B.3	Model and record 2-digit addition.	R—4.5	2.NBT.6
5, 6	4.3	CC.2.1.2.B.3	Add 2-digit numbers using mental math.	R—4.3	2.NBT.5
19, 20	4.9	CC.2.2.2.A.2	Add 2-digit numbers.	R—4.9	2.OA.1
11, 12, 25	4.12	CC.2.1.2.B.3	Add 4 numbers.	R—4.12	2.NBT.6
13, 14	4.7	CC.2.1.2.B.3	Decide whether a sum is greater than or less than 100.	R—4.7	2.NBT.5
17, 18	4.1	CC.2.1.2.B.3	Find equivalent ways to write a sum.	R—4.1	2.NBT.6
1, 2	4.2	CC.2.1.2.B.3	Use compensation to develop flexible thinking for 2-digit addition.	R—4.2	2.NBT.5
7, 8	4.4	CC.2.1.2.B.3	Model 2-digit addition with regrouping.	R—4.4	2.NBT.6
15, 16	4.8	CC.2.1.2.B.3	Rewrite horizontal addition problems vertically in the standard algorithm format.	R—4.8	2.NBT.6

Key: R—Reteach

Assessment Guide
© Houghton Mifflin Harcourt Publishing Company

Individual Record Form

Child's Name _____ Date _____

Chapter 5 Test

Item	Lesson	PA Common Core Standard	Content Focus	Intervene With	Personal Math Trainer
6, 19	5.5	CC.2.1.2.B.3	Determine whether regrouping is necessary when subtracting.	R—5.5	2.NBT.5
4, 16	5.8	CC.2.1.2.B.3	Use a number line to add to find a difference.	R—5.8	2.NBT.5
9, 24, 25	5.9	CC.2.2.2.A.2	Use a bar model to solve a subtraction problem.	R—5.9	2.OA.1
3, 15	5.1	CC.2.1.2.B.3	Break apart ones to subtract.	R—5.1	2.NBT.5
8, 21	5.2	CC.2.1.2.B.3	Break apart tens and ones to subtract.	R—5.2	2.NBT.5
2, 14	5.7	CC.2.1.2.B.3	Rewrite and subtract.	R—5.7	2.NBT.5
10, 11, 22	5.11	CC.2.2.2.A.2	Complete bar models to solve a multistep subtraction problem.	R—5.11	2.OA.1
7, 20	5.4	CC.2.1.2.B.3	Record subtraction with regrouping.	R—5.4	2.NBT.5
1, 13	5.3	CC.2.1.2.B.3	Model regrouping in subtraction.	R—5.3	2.NBT.5
12, 23	5.10	CC.2.2.2.A.2	Write an equation to solve a subtraction problem.	R—5.10	2.OA.1
5, 17, 18	5.6	CC.2.1.2.B.3	Practice 2-digit subtraction with and without regrouping.	R—5.6	2.NBT.5

Key: R—Reteach

Assessment Guide
© Houghton Mifflin Harcourt Publishing Company

AG132

Individual Record Form

Child's Name _____ Date _____

Chapter 6 Test

Item	Lesson	PA Common Core Standard	Content Focus	Intervene With	Personal Math Trainer
7, 8, 16	6.3	CC.2.1.2.B.3	Regroup ones to add.	R—6.3	2.NBT.7
10, 21, 22	6.6	CC.2.1.2.B.3	Subtract 3-digit numbers without regrouping.	R—6.6	2.NBT.7
6, 18, 25	6.2	CC.2.1.2.B.3	Break apart 3-digit addends.	R—6.2	2.NBT.7
5, 17	6.1	CC.2.1.2.B.3	Draw to represent 3-digit addition.	R—6.1	2.NBT.7
11, 23	6.9	CC.2.1.2.B.3	Regroup hundreds and tens to subtract.	R—6.9	2.NBT.7
3	6.8	CC.2.1.2.B.3	Regroup hundreds to subtract.	R—6.8	2.NBT.7
9, 19, 20	6.5	CC.2.1.2.B.3	Regroup ones and tens to add.	R—6.5	2.NBT.7
4, 14, 15	6.4	CC.2.1.2.B.3	Record 3-digit addition using the standard algorithm with possible regrouping of tens.	R—6.4	2.NBT.7
1, 2, 13	6.7	CC.2.1.2.B.3	Record 3-digit subtraction using the standard algorithm with possible regrouping of tens.	R—6.7	2.NBT.7
12, 24	6.10	CC.2.1.2.B.3	Record subtraction using the standard algorithm when there are zeros in the minuend.	R—6.10	2.NBT.7

Key: R—Reteach

Assessment Guide
© Houghton Mifflin Harcourt Publishing Company

Individual Record Form

Child's Name _____   Date _____

Chapter 7 Test

Item	Lesson	PA Common Core Standard	Content Focus	Intervene With	Personal Math Trainer
23	7.7	CC.2.4.2.A.3	Select combinations of bills and coins with a given value.	R—7.7	2.MD.8
12, 24	7.11	CC.2.4.2.A.2	Tell time from a clock as A.M. or P.M. based on the problem scenario.	R—7.11	2.MD.7
7	7.4	CC.2.4.2.A.3	Show an amount using coins.	R—7.4	2.MD.8
11, 21	7.5	CC.2.4.2.A.3	Count a collection of coins and compare to $1.00.	R—7.5	2.MD.8
3, 4, 15	7.9	CC.2.4.2.A.2	Tell time to the nearest 5 minutes.	R—7.9	2.MD.7
9, 19	7.10	CC.2.4.2.A.2	Tell time as minutes after an hour.	R—7.10	2.MD.7
5, 22	7.6	CC.2.4.2.A.3	Count a collection of a bill and coins with a total greater than $1.00.	R—7.6	2.MD.8
8, 17, 18, 25	7.2	CC.2.4.2.A.3	Count a collection of coins.	R—7.2	2.MD.8
1, 2, 13, 14	7.8	CC.2.4.2.A.2	Tell time to the half hour.	R—7.8	2.MD.7
6, 16	7.1	CC.2.4.2.A.3	Count a collection of coins.	R—7.1	2.MD.8
10, 20	7.3	CC.2.4.2.A.3	Order coins in a collection by value and then find the total value.	R—7.3	2.MD.8

Key: R—Reteach

Assessment Guide
© Houghton Mifflin Harcourt Publishing Company

Individual Record Form

Child's Name _____ Date _____

Chapter 8 Test

Item	Lesson	PA Common Core Standard	Content Focus	Intervene With	Personal Math Trainer
4, 15	8.8	CC.2.4.2.A.1	Choose a tool and explain.	R—8.8	2.MD.1
6, 17, 23	8.5	CC.2.4.2.A.6	Relate addition to length and use a number line diagram.	R—8.5	2.MD.5, 2.MD.6
8, 9, 19	8.4	CC.2.4.2.A.1	Use a ruler to measure an object.	R—8.4	2.MD.2
5, 16, 20	8.1	CC.2.4.2.A.1	Measure length with an inch model.	R—8.1	2.MD.1
1, 2, 12, 13, 23	8.9	CC.2.4.2.A.4	Use a line plot and explain how it will change with data changes.	R—8.9	2.MD.9
3, 14	8.6	CC.2.4.2.A.1	Select inches or feet as the correct units for given measures.	R—8.6	2.MD.2
7, 18	8.3	CC.2.4.2.A.1	Estimate length using an inch model.	R—8.3	2.MD.3
10	8.7	CC.2.4.2.A.1	Estimate length in feet.	R—8.7	2.MD.3
11, 21, 22	8.2	CC.2.4.2.A.1	Make an inch ruler and use it to measure the lengths of objects.	R—8.2	2.MD.3

Key: R—Reteach

Child's Name _____ Date _____

Chapter 9 Test

Item	Lesson	PA Common Core Standard	Content Focus	Intervene With	Personal Math Trainer
2, 3, 12	9.1	CC.2.4.2.A.1	Measure length using a centimeter model.	R—9.1	2.MD.1
7, 8, 10, 17	9.2	CC.2.4.2.A.1	Estimate length in centimeters.	R—9.2	2.MD.3
5, 14, 15, 20	9.4	CC.2.4.2.A.1	Relate addition to length and use a number line diagram.	R—9.4	2.MD.4
1	9.5	CC.2.4.2.A.1	Select centimeters or meters as the correct unit for given measures.	R—9.5	2.MD.1
4, 13	9.6	CC.2.4.2.A.1	Estimate length in meters.	R—9.6	2.MD.3
6, 11, 16	9.7	CC.2.4.2.A.1	Measure and compare lengths of two objects.	R—9.7	2.MD.4
9, 18, 19, 20	9.3	CC.2.4.2.A.1	Measure length to the nearest centimeter.	R—9.3	2.MD.1

Key: R—Reteach

Assessment Guide
© Houghton Mifflin Harcourt Publishing Company

AG136

Individual Record Form

Child's Name _____ Date _____

Chapter 10 Test

Item	Lesson	PA Common Core Standard	Content Focus	Intervene With	Personal Math Trainer
9–12, 21–24	10.1	CC.2.4.2.A.4	Make a tally chart.	R—10.1	2.MD.10
1, 2, 14	10.5	CC.2.4.2.A.4	Make and interpret a bar graph.	R—10.5	2.MD.10
3, 4, 15, 16	10.4	CC.2.4.2.A.4	Read and interpret a bar graph.	R—10.4	2.MD.10
7, 8, 19, 20	10.3	CC.2.4.2.A.4	Complete a picture graph.	R—10.3	2.MD.10
5, 6, 17, 18, 25	10.2	CC.2.4.2.A.4	Read and interpret a picture graph.	R—10.2	2.MD.10

Key: R—Reteach

Assessment Guide
© Houghton Mifflin Harcourt Publishing Company

Individual Record Form

Child's Name _____   Date _____

Chapter 11 Test

Item	Lesson	PA Common Core Standard	Content Focus	Intervene With	Personal Math Trainer
7, 8, 19, 20	11.1	CC.2.3.2.A.1	Match objects and three-dimensional shapes.	R—11.1	2.G.1
2, 11, 23	11.2	CC.2.3.2.A.1	Identify attributes of a rectangular prism.	R—11.2	2.G.1
12, 24	11.8	CC.2.3.2.A.2	Draw to show equal parts of a two-dimensional shape.	R—11.8	2.G.3
6, 18	11.10	CC.2.3.2.A.2	Draw halves, thirds, and fourths.	R—11.10	2.G.3
9, 21	11.7	CC.2.3.2.A.1	Identify and name equal parts.	R—11.7	2.G.2
1, 13, 14, 25	11.3	CC.2.3.2.A.1	Determine the number of cubes in a rectangular prism.	R—11.3	2.G.1
5, 17, 25	11.4	CC.2.3.2.A.1	Count sides of two-dimensional shapes to solve a problem.	R—11.4	2.G.1
3, 15	11.5	CC.2.3.2.A.1	Draw two-dimensional shapes with a given number of angles.	R—11.5	2.G.1
10, 22	11.9	CC.2.3.2.A.2	Identify shapes divided into thirds.	R—11.9	2.G.3
4, 16	11.6	CC.2.3.2.A.1	Partition rectangles into equal-size squares and find the total number of these squares.	R—11.6	2.G.1

Key: R—Reteach